The Sierra Club Handbook of
Whales and Dolphins

The Sierra Club Handbook of

WHALES AND DOLPHINS

Stephen Leatherwood and
Randall R. Reeves
Paintings by Larry Foster

SIERRA CLUB BOOKS   SAN FRANCISCO

COPYRIGHT 1983 in all countries of the International Copyright Union by Stephen Leatherwood, Randall R. Reeves, and Larry Foster.

All rights reserved. No part of this book may be reproduced in any form or by any electronic or mechanical means, including information storage and retrieval systems, without permission in writing from the publisher.

Library of Congress Cataloging in Publication Data

Leatherwood, Stephen.
    The Sierra Club Handbook of Whales and Dolphins.

    Bibilography: p.
    Includes index.
    1. Cetacea. I. Reeves, Randall R.   II. Foster, Larry,
1909–   III. Title
QL737.C4L33 1983   599.5   83-388

ISBN: 0-87156-341-X
ISBN: 0-87156-340-1 (pbk.)

Book and jacket design by Jon Goodchild
Printed by Dai Nippon Printing Company, Ltd.,
Tokyo, Japan.

10   9   8   7   6   5   4   3

For Aubrey, Kathleen, Keith,
and Shannon Leatherwood
and
For Harold and Bernice Reeves
and
For Del and Anita Foster and
Nick and Muriel Tannaci

Contents

Foreword

THIS BOOK IS IDEAL for those who want to go out to see whales and recognize what they have seen. The authors, Stephen Leatherwood and Randall Reeves, know whales large and small. They have a balanced familiarity with the creatures viewed from every angle: from ships plying the sea, from aircraft flying low over water, from the depths through divers' faceplates, and through windows of oceanarium tanks. Unlike many popular accounts of cetaceans, this volume is enriched by personal encounters and first-hand observations.

Having taken a stand solidly on the side of whales, these authors retain a protectionist viewpoint; yet their scholarship, objectivity, and observational abilities prevail. They combine a genuine admiration and respect for the animals with the research biologist's eye for factual, accurate detail. For those whose knowledge of whales comes mainly from television and newspapers, this book is therefore an excellent place to begin learning. The facts, as known in the early 1980s, are well presented here.

Fifteen years ago Stephen Leatherwood and I tumbled out of a hovering helicopter to scuba through the kelp and sea lions of San Nicholas Island; since then we have shared several adventures while observing marine mammals. My acquaintance with Randall Reeves has resulted from our involvement with the efforts of the U.S. Marine Mammal Commission. I know each man to be diverse in experience and rounded in knowledge. Both obtained bachelor's degrees in liberal arts before migrating toward a biological interest in whales. Perhaps their training in the humanities explains their ability to write engagingly about whales, describing them in a way that does justice to the sense of wonder they arouse without adding flummery and mysticism to the subject.

For a book such as this, good writing is not enough. The authors also had the sense to join with talented artist/research team Larry and Valerie Foster, who have learned to portray the details, bringing life to whales on canvass perhaps better than anyone ever has. Getting these illustrations together under one cover is itself adequate justification for this book.

SAM H. RIDGWAY, D.V.M., Ph.D.
San Diego, California

Preface

WHEN IT WAS FIRST SUGGESTED to us early in 1978 that we write a handbook on cetaceans of the world, we were reluctant, though clearly not because we lacked enthusiasm for the subject. Together we can claim over twenty years of experience in studying, lecturing, and writing about marine mammals. Not only our careers, but often our personal lives as well, have become bound up one way or another with sea mammals. Furthermore, we both came to this field of study from different careers, drawn in part by the mystique that has made marine mammals, especially cetaceans, among the most provocative and admired of all creatures. Over the years and throughout our exposure to the whales, dolphins, and porpoises, we have become more curious about their lives and more concerned about their conservation.

We balked at the thought of writing a book quite simply because there have been so many recent books on marine mammals that we thought we would be cluttering an already overloaded market. But as we examined the book shelves and discussed our colleagues' books, we saw a pattern. On the one hand, there were the technical volumes; however precise and up-to-date the information in them may be, they are often so full of technical detail and jargon they are unreadable for most people. On the other hand, there were the popularized accounts; most of these offer nothing new, but instead continue to parrot myths borrowed repeatedly from surprisingly few original sources. As one author recently pointed out, there is a serious danger in the literature on marine mammals that certain notions will come to be regarded as true simply because they have been repeated so many times. These popular books often fail to acknowledge what to our minds is obvious, that even when the myths are stripped away and replaced by a few unembellished facts, cetaceans are truly remarkable creatures.

We found in the literature little to bridge the gap between these extremes, and were convinced of the need for a handbook written in plain language, aimed at a broad, nontechnical audience, which respectfully kept pace, as well as any summary can, with the current status of a fast-moving science. But this conviction alone would not have moved us to write this book. We remembered George Orwell's admonition that anyone who would undertake to write a book is possessed of an evil demon, and we had other demons in possession of much of our energy.

Our decision was made when we learned that an extraordinary artist, Larry Foster, was willing, even enthusiastic, to cast his lot in with us. We considered it an honor to have our words provide a gallery for his art and scholarship, for he is both a student concerned with the forms and dimensions of cetaceans, and an artist keenly attuned to their beauty. So the three of us reached agreement and began our tasks, and by the autumn of 1979 had finished the manuscript and paintings. Problems that do not merit discussion threatened to prevent publication of this book; so we were much relieved and elated when the book was rescued by the interest and support of the Sierra Club.

We sincerely hope that what we offer here is a true book, fair to the best of what we know about the animals' natural history and appearance. We have taken great pains to make it right, and have shamelessly taken advantage of our colleagues' willingness to read the manuscript, catch errors, and offer new information. In a field such as this, depending only on the published literature assures inadequacy, since so much of what is known has not yet reached print, and even with the generosity of so many associates, what is known will far exceed what we have said of it even before it reaches you, the reader.

We have a profound respect for the immutability of the printed word, and are aware of how colorlessly it sometimes portrays nature. But in the text we have written and in what Larry has painted, we hope we have not lost sight of the fascinating animals that inspired this book, and we hope it will be a useful, accurate, and attractive reference.

We owe thanks to a great many people, without whose help and encouragement this book would never have been possible. Our greatest debt is to Valerie Cross Foster, the backbone and heart of General Whale, who informed and inspired us during several long sessions of firsthand collaboration, and kept the work flowing efficiently throughout the project. We would have found the task perhaps impossible, and certainly much less pleasant, without her tireless help.

The photographers whose contributions enliven the pages of this book have been very generous (credits accompany each photograph). We are indebted to them not only for the credited photographs used in the book, but also for their contributions to the General Whale photographic files.

We and Sierra Club Books especially thank the following individuals, who ensured technical accuracy by reviewing major portions of this manuscript: Robert L. Brownell, Jr., Giuseppe Notarbartolo di Sciara, James G. Mead, William F. Perrin, Dale W. Rice, Sam H. Ridgway, and Forrest G. Wood. We also must thank the following individuals for reviewing selected species accounts: Alan N. Baker, R. Natalie P. Goodall, John D. Hall, Larry J. Hobbs,

Mary Lou Jones, Steven K. Katona, Didier Marchessaux, Charles W. Potter, Graham J. B. Ross, Brent S. Stewart, Steven L. Swartz, William A. Walker, Randall S. Wells, Bernd Würsig, and Pamela Yochem. Their criticisms have improved the manuscript greatly. The errors that remain are ours alone.

Elizabeth Mooney, Derrith Bartley, and Randi Olsen assisted with literature review, photograph sorting and cataloging, and preparation of draft maps. Stan Minasian of the Marine Mammal Fund generously shared his collection of cetacean photographs, particularly those of the more poorly documented species. Maeve K. Edwards and Angela Rowley typed the manuscript. Aidan A. Kelly, Robert Buhr, and Judy Rand edited the manuscript.

Finally, for the generous spirit in which they have tutored and encouraged us over the years, we especially thank the late Carl L. Hubbs (who died in 1979), Clayton E. Ray, William F. Perrin, James G. Mead, Edward D. Mitchell, Robert L. Brownell, Jr., Sam H. Ridgway, Forrest G. Wood, and William E. Evans.

STEPHEN LEATHERWOOD
RANDALL REEVES
San Diego, California
January 1982

Notes From the Artist
and the Researcher

ONE OF THE VALUES OF PAINTED ILLUSTRATION is in the artist's option to idealize a subject. This option has been exercised in these water color (tempera) cetaceans, which are idealized to give a very clear idea of the specific nature of the various details, large and small, of the external anatomy of each species. The paintings' subjects are realistic rather than naturalistic. Flippers tend to be all perpendicular views, and tail flukes are tipped away from horizontal just enough to give some idea of their specific shapes. Water serves as decorative background only, with none appearing in front of the animal. Elements not essential to species identification have been left out, such as teeth scratches, nicks on dorsal fins, certain external parasites, and so on. The smallest cetacean species is the smallest in printed size in these illustrations, and the largest species, the largest printed size. The printed sizes of all other species fall in between in relation to the size of the actual animal. But this scale is not exactly true to nature. In life, the length of the blue whale's flipper is over twice the total body length of the smallest river dolphin.

The sophisticated quality of these "anatomical maps" is partially due to the steady brush of the craftsman. However, to produce such illustrations, much *more* effort must be given to research. The likelihood of the whale illustrator ever actually seeing the subjects alive and wild in the sea is remote. Most information about the shape and size of whales comes to us through published and unpublished measurements and photographs. The noting and collecting of this data is the main activity of General Whale, our non-profit research organization. Fortunately, many people are now looking at living whales with cameras, as the photographs we have included herein show.

But our fluked friends in the sea remain elusive and inconvenient subjects to study. In order to acquire a more thorough understanding of the physical details of the various whale, dolphin, and porpoise species, much more work must be done by photographers and artists. In these illustrations can be seen some of the progress made so far.

LARRY AND VALERIE FOSTER
Alameda, California
November 1982

The Sierra Club Handbook of
Whales and Dolphins

Conversion Factors

TO CHANGE	INTO	MULTIPLY BY
Centimeters	Inches	0.3937
Meters	Feet	3.281
Meters	Yards	1.094
Kilometers	Statute miles	0.6214
Kilometers	Nautical miles	0.5396
Nautical miles	Kilometers	1.853
Square kilometers	Square nautical miles	0.29
Square nautical miles	Square kilometers	3.43
Fathoms	Feet	6.0
Fathoms	Meters	1.829
Kilograms	Pounds	2.205
°Centigrade	°Fahrenheit	1.8, then add 32

Introduction to the *Order* Cetacea:

WHALES, DOLPHINS & PORPOISES

Introduction to the Cetaceans

We do not expect this volume to be, for most of its readers, a first encounter with the cetaceans (whales, dolphins, and porpoises), since so much material—books, articles, films, artwork—about cetaceans is now available. Most readers have probably at least seen dolphins and whales on television; others have visited oceanaria to see live bottlenose dolphins, killer whales, white whales, or a variety of less frequently displayed species; and those in a luckier minority have seen the animals in the wild, frolicking in a ship's bow wave, or breaching and spouting on the horizon. Because of this exposure, we expect much of the reading public will be more aware of these animals than of many wild creatures. But to refresh the memory and to reconstruct the context, and perhaps genuinely to introduce some readers to the "spouting fish," as Melville dubbed the cetaceans, we shall reiterate what are, to our minds, the bare essentials.

CLASSIFICATION AND NOMENCLATURE: The order Cetacea (from the Greek word *ketos* and the Latin word *cetus,* both meaning "whale") includes three suborders: THE ARCHAEOCETI or "ancient whales," extinct forms known only from fossils; THE MYSTICETI or "moustached whales," including at least ten living species of baleen, or whalebone, whales; and THE ODONTOCETI or "toothed whales," including 65 or more living species of dolphins, porpoises, and whales with teeth but no baleen.

In addition to a variety of other anatomical, physiological, and ecological differences, baleen and toothed whales differ in two very striking ways:

(1) The Mysticeti have baleen; the Odontoceti have teeth.

(2) The Mysticeti have two external blowholes; the Odontoceti have one.

Instead of teeth, baleen whales have a series of plates, constructed of material much like human fingernails, rooted in the roof of the mouth. Each of the plates forms roughly a right triangle, with the long vertical edge oriented to face outside the mouth, the base embedded in the gum, and the hypotenuse inside the mouth. The outer edge is smooth; the inner edge is frayed, and consists of a series of bristles or "hairs," which differ in appearance and number from one species to another, and can be used to distinguish them. The plates are arranged somewhat like the teeth in a comb, one row

The capacious mouth of a large bowhead whale, like that of this 16.5-m male killed at Barrow, Alaska, can contain supple strips of baleen up to 4 m long. (August 28, 1976: Les Nakashima, courtesy of U.S. Naval Arctic Research Laboratory.)

along each side of the mouth. Inside the mouth the bristles intertwine to form a sieve or mat.

Baleen whales feed in at least three ways, depending on the species:

(1) by opening the mouth wide, taking in a large quantity of water, and then closing the mouth, forcing the water out and collecting on the mat of bristles the plankton or fish the water contains;

(2) by swimming through the water, mouth ajar, allowing the water to flow in through the gap at the front of the mouth between the two rows of baleen plates, and out through the gaps between the side plates, leaving the particles of food behind;

(3) by using the suction created when the tongue is pressed against the palate to draw water and food into the mouth, then squeezing the water out with another movement of the tongue.

Food of baleen whales ranges from krill (small shrimplike crustaceans) and other zooplankton up to schooling fishes. The structure of the mat formed by the baleen—the size and flexibility of the individual bristles, and the density in which they normally occur—is related to the food preference of each species. Those that feed on small planktonic organisms tend to have finer, more closely packed fringes than those whose diet includes fish.

Unlike the baleen whales, toothed whales do have teeth, though these teeth may erupt only in males (most beaked whales and the narwhal), take a curious shape (the tusks of narwhals and certain mesoplodonts), and become badly worn or completely shed in older individuals (Risso's dolphins, bottlenose dolphins, and white whales). As with the baleen of mysticetes, the dentition of odontocetes is roughly related to their feeding habits: those species that feed primarily on squid have fewer or no visible teeth; those with more varied diets, particularly those whose diets include small schooling fish, generally have a longer snout and many teeth.

Baleen whales have two external blowholes that generate two columns of vapor, although in most species the two columns merge to create a single spout. Toothed whales, on the other hand, have only one blowhole. Even though internally they, like the baleen

The sperm whale's long, narrow lower jaw carries a formidable array of conical teeth. (Azores, 1979: courtesy of Jonathan Gordon.)

The two blowholes of a southern right whale produce a bushy, V-shaped blast. (Mother and calf off Geraldton, Western Australia: Patrick Baker.)

The sperm whale's single blowhole is set far forward and to the left, giving its blow an oblique trajectory. (Near Pennell Bank, Antarctica, 72°48'S, 179°51'E, January 27, 1981: Gerry Joyce.)

whales, have two openings to the nasal passage divided by a central septum, their septum is internal; so only one blowhole is visible externally. When an odontocete blow is visible at all, it consists of a single column or puff of vapor.

Included among the baleen whales are the majority of the large species, from the awesome blue whale, whose maximum length and weight may never have been accurately measured but surely exceeds 30 m and 160 tons,* to the Bryde's whale, at 14 m a little less than half the length of the blue. The abundant minke whale (found in all oceans) and the rare pygmy right whale (reported only from

* See note on p.22

the southern hemisphere) are the diminutive members of the sub-order, reaching lengths of little more than 10 m and 7 m, respectively. Baleen whales are widely distributed, with one form or another present at least seasonally in almost every major ocean and sea in the world. As a group, the baleen whales are usually divided into three families: the Balaenidae (right whales) with no dorsal fin, no ventral grooves or throat creases, and long, narrow baleen plates; the Eschrichtiidae (gray whales), with two to five throat creases, and a small dorsal hump followed by serrations; and the Balaenopteridae (rorqual whales) with numerous ventral grooves and a dorsal fin. The pygmy right whale, classified as a balaenid, has a dorsal fin but no ventral grooves, and in many other respects it is intermediate between the right whales and the rorquals.

The range in size among toothed whales is even wider, from the mighty male sperm whale, which at 18 m is fully as long as many of the large baleen whales, to the diminutive coastal porpoises (*Phocoena* spp. and *Cephalorhynchus* spp.), which rarely reach 2 m in length. The toothed whales are currently sorted into six families: Physeteridae (sperm, pygmy sperm, and dwarf sperm whales); Monodontidae (white whales and narwhals); Ziphiidae (beaked whales); Delphinidae (dolphins); Phocoenidae (true porpoises); and Platanistidae (freshwater river dolphins and the franciscana).

Taxonomy, the science of classifying animals, is a controversial aspect of cetology, regarded by some as vital and exciting, by others as a distasteful but necessary form of bookkeeping. We indulge in classification in this book for its convenience, ever mindful that many of the relationships attributed to the animals are not well-established, let alone "proven." For example, a captive bottlenose dolphin at Sea Life Park in Hawaii bred successfully with a rough-toothed dolphin, a fact that calls into question the generic separation of these two species. (Some taxonomists still put them in different families, Delphinidae and Stenidae.) More recently, captive bottlenose dolphins have bred successfully with Risso's dolphins and false killer whales in Japan, and with a short-finned pilot whale in California (although the calf from this last mating was stillborn). So far, none of the hybrids produced by these crosses has lived long enough to test their reproductive viability. The late F. C. Fraser of the British Museum once described a group of wild dolphins stranded on an Irish beach as possible hybrids, with attributes of both bottlenose dolphins and Risso's dolphins. Morphological intergradation, possibly resulting from interbreeding between species in the wild, has been noted among several of the *Balaenoptera* species as well.

It is still possible to discover new species of cetaceans or radically reinterpret the relationships between currently recognized species. The beaked whales, in particular, continue to offer major challenges to systematists. One species, *Mesoplodon pacificus*, has

A hybrid calf, born May 3, 1981, shown here with its bottlenose dolphin mother (left) and its father, a false killer whale (top). (Kamogawa Sea World, Japan, 1981: Masaharu Nishiwaki.)

never been seen and identified alive or dead; it is known only from two skulls recovered from beaches. There are bottlenose whales in the equatorial Pacific, first reported by Ken Balcomb, and subsequently seen by observers on American tuna boats; in the complete absence of specimens, the identity and affinities of these whales remain conjectural.

The mysterious process by which animals are classified and named has taxed our imagination and patience during certain phases of this book's preparation. Because of our own interest, we have tried to explain for readers the etymology of the scientific names of all the cetacean species. Now we know why no one had carried out such an exercise. Authors have rarely explained in print the basis for the names they have given to the cetaceans. Since we are not expert in Greek or Latin, we have pondered lengthily, but not always fruitfully, the etymological possibilities. We are not entirely satisfied with the results, and can only claim that they represent our best effort, nothing more.

Decisions about how many genera and species there are, what author deserves credit for naming them, and which names should be given precedence, ought not to be made lightly. However, this book, written for a popular audience, is not the place to argue for new systematic relationships or to change conventional nomenclature. Therefore, we agreed from the outset to follow the lists of

The dusky dolphin of the southern oceans (above) is almost identical to the Pacific white-sided dolphin (page 9), a North Pacific endemic. They are assigned to separate species because of disjunct distribution and slight morphological differences. (Above, southeastern Golfo San Jose, Argentina, December 1974: © Mel and Bernd Würsig. Page 9, off central California, 37° 14' N, 123° 07' W, June 24, 1980: Robert L. Pitman.)

names developed by the Scientific Committee of the International Whaling Commission, which we consider the only authoritative international body concerned with cetacean management and biology, and by the U.S. Marine Mammal Commission.*

Yet another problem was the selection of common names. Many have been coined by zoologists with little or no first-hand familiarity with the animals; others, by local natives whose intimate personal acquaintance with the animals has led to colorful and highly appropriate names that reflect characteristic features of the species. Rather than trying to track down every common name applied historically or geographically to every species, or deciding arbitrarily to list some and ignore others, we finally chose to adhere strictly to the lists of English common names used by the Whaling Commission and the Marine Mammal Commission. However flawed they may be, these lists are as close to a consensus as is now available.

In several instances we have lumped together two recognized species (e.g., the Ganges and Indus susus, the northern and southern right whales), treating them in the illustration and narrative as though there were no significant differences between them. Our intention was not to cloud further an already fuzzy picture. We are quite simply hamstrung by our ignorance. Larry said he was unable to distinguish them accurately with his brush and canvas, and for many of these pairs too little is known about the biology and behavior of one or both species to enable separate accounts to stand on their own.

---

* The names we have used for the large whales follow longstanding usage by the International Whaling Commission (IWC), except that the name for the sperm whale used by the IWC has recently been changed from *Physeter catodon* to *P. macrocephalus*. The scientific and common names we have used for the smaller cetaceans follow a list published in 1977 by the IWC.

EVOLUTION: The fossil history of cetaceans is being pieced together by paleontologists, relying so far on scattered fragments excavated from deposits in exposed seabeds. In general, the mosaic consists of a few lonely tiles separated by broad gaps of ignorance. The uncertainty becomes progressively greater as one moves farther into the past. Some relatively modern forms have been well-described, but many intermediate and earlier forms are still undiscovered. Not surprisingly, then, much of what has been said and written about cetacean origins is speculative, and where there is room for speculation, there is at least equal room for disagreement.

We do know that essentially modern forms of baleen and toothed whales were present by 10 or 12 million years ago in the Late Miocene and that forms closely resembling modern cetaceans were present at least by the Early Miocene, about 23 million years ago. Some of the latter, for example the Cetotheriidae, proliferated in the Early Miocene and became extinct millions of years later in the Pliocene. They gave rise during the Miocene to the baleen whales we know today. The Archaroceti, extinct whale ancestors, flourished in the Eocene, about 45 to 55 million years ago. The question of whether these archaeocetes gave rise to the two extant groups of cetaceans is a matter of scholarly contention.

Controversy has long existed, for example, about whether the two major extant suborders (Mysticeti and Odontoceti) arose from a common ancestral group or from two separate sources, each molded into similiar form and function by the same environmental exigencies. Were the similarities of two different groups highlighted by evolution, forcing them to converge; or did the evolutionary process exploit differences within a single group of organisms, allowing it to diverge and produce the variety of forms known today? The debate wears on in paleontological circles, informed but unresolved by such modern techniques as genetic karyotyping, myoglobin sequencing, and electrophoresis (protein analysis). Whatever their number and appearance, the remote ancestors of cetaceans are thought to have ventured back into the water about 55 million years ago.

ADAPTATIONS: Over the centuries, cetaceans as a group have managed to exploit virtually all types of productive marine, estuarine, and major riverine habitats. A few species have made a permanent home among polar ice floes; others live out their lives in freshwater channels that wind through tropical rain forests. Cetaceans can be found from muddy, coastal shallows to crystal-clear oceanic regions. Some (e.g., blue, beaked, and pygmy and dwarf sperm whales) occur individually or in small groups; others (e.g., the killer whale) in slightly larger groups, which appear to remain together for many generations; and still others—primarily the oceanic dolphins—in enormous assemblages made up of many smaller functional units, creating what Melville called "hilarious shoals, which

Nothing enlivens the seascape, or lifts the seafarer's spirits, like a shoal of dolphins, in this instance Delphinus delphis. *(Off southern California, 32° 30' N, 117° 19' W, March 17, 1971: Stephen Leatherwood.)*

upon the sea keep tossing themselves to heaven like caps in a Fourth-of-July crowd."

The variety of their habitats and their systems of organization testify to the completeness of the modern cetaceans' invasion of the water world. They are supremely adapted to marine life as mammals. Short of their becoming fish, it is difficult to imagine how they could better meet the demands of their new medium, which is assumed to be basically hostile to mammalian life. They must propel themselves through water at speeds and to depths sufficient for overtaking prey and avoiding predators, breathe air (in the presence of water) and hold their breath for protracted periods, maintain a fairly constant core body temperature in an environment that dissipates heat much more rapidly than air, and bear and suckle their young in the water. They must live under circumstances that often render the mammalian senses, as we know them from our terrestrial experience, useless in meeting the daily challenges of finding food, avoiding enemies, caring for young, and tending other social bonds vital to survival. We have a profound respect for the sometimes sweeping, sometimes subtle ways in which cetaceans have solved these problems. We are able here only to touch on some of the highlights of their adaptation to a marine existence.

Cetaceans are streamlined. The breathing mechanism (one or two blowholes) has migrated to the top of the head, accompanied by a pronounced telescoping of the skull in which both the upper and the lower jaws have extended far forward of the bony entrance to the nares. With the blowholes in this position, the cetaceans can exchange air without interrupting the smooth forward motion of the body. Drag is reduced by internalization of certain body parts: the teats and genitals are concealed in slits in the body wall, and the external ear has disappeared, leaving only a pinhole opening or a membrane flush with the skin. The two forelimbs have become flattened, paddlelike flippers, the "hand" and "finger" bones being

A diver's view of bottlenose dolphins. (Off Houtman's Abrolhos Island, Western Australia, 1981: Patrick Baker.)

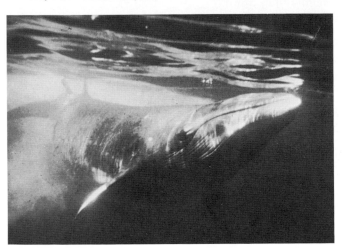

Despite the harpoon in its flank, this 4.9-m minke whale demonstrates the graceful body shape of the living rorquals. (About 20 to 40 km off the Pacific coast of Japan, July 1970: Gordon Williamson.)

contained within a common integument. Each flipper is usually rotated as a unit at the shoulder blade and used for stability and steering. At times the flippers are pressed against the body; some beaked whales even have slight indentations on the sides into which the flippers fit snugly (so-called "flipper pockets").

In most species some sort of irregularity—usually a fin, but sometimes a hump or ridge—is present along the midline of the back. It could act as a kind of keel, or perhaps it is more important as a medium for heat exchange. However, species that lack a dorsal fin appear equally suited to their roles. Bowhead whales make their way among the heavy ice floes of the Arctic, and right whale dolphins race across vast oceanic expanses in temperate latitudes, both apparently finding no need whatsoever for a dorsal fin.

There is no external evidence of hind limbs on cetaceans; small bones embedded in the side muscle are the rudiments of the pelvic girdle and sometimes of the femur. In rare instances these have been found to support small, rudimentary external limbs, betraying the terrestrial origins of cetaceans. These bones serve in some species as a point of origin or insertion for muscles, but are otherwise vestigial.

The fibrous, horizontally flattened tail flukes have no skeletal support. The rear third of the body is really a powerful, heavily muscled tail, often called the tail stock or caudal peduncle. It is laterally compressed, probably to reduce drag during the up-and-down swimming motions (and perhaps to aid the dorsal fin by providing an added measure of stability). It drives the flukes, which in

Rough-toothed dolphins seen from the bow-bubble of the research vessel David Starr Jordan. *(Southwest of Acapulco, Mexico, 14°08′ N, 104°38′ W, February 25, 1979: Robert L. Pitman.)*

turn provide the propulsion.

Cetaceans have almost, but not entirely, lost the hairy external covering characteristic of other mammals. Hairs can still be seen along the snout of younger animals and in specialized locations on the head of adults of some species. The skin is smooth and resilient. It has been suggested that a specialized pattern of water flow along the body helps minimize drag and increase swimming efficiency in some well-studied dolphin species. Some investigators have argued that dolphins can swim faster than theoretical "hull" speeds. Furthermore, recent underwater observations of dolphins swimming at high speed have inspired speculation that at least some species may use a muscularly controlled "shiver" to dump excess drag and to establish the special characteristics of water flow along the body

Underwater view of a gray whale's flukes, in the kelp beds off Point Loma. (San Diego, California, January 1970: Howard Hall.)

necessary for efficient swimming.

It had long been speculated that the ventral grooves of balaen-opterid whales were somehow hydrodynamically important to these fast-swimming species. The grooves may make the animal's throat more compressible by pressure during deep dives and by muscular contraction during surface swimming. But recent observations of feeding balaenopterids of five different species, in which the jaws are opened wide and the throat distended into grotesque guppy-like shapes, demonstrate conclusively that the ventral grooves play some important role in the feeding process.

Although ultimately tied to the surface for life-giving air, cetaceans have developed the ability to explore and exploit the ocean's depths. A wild common dolphin, which was captured, radio-tagged with special instrumentation, released, and tracked, dived to a depth of 260 m, remaining submerged for more than eight minutes on one occasion, and reportedly exceeded 90 m and three minutes routinely. A similarly instrumented pilot whale dived to over 500 m, and once may have reached 600 m; its dives lasted more than 16 minutes. Sperm whales have been found entangled in submarine cables at depths of more than 1,000 m, and have been tracked by active sonar and hydrophones to depths of 2,000 and 2,800 m, respectively, during dives lasting an hour or more. The depths to which bottlenose whales (the apparent diving champions, with dives reportedly lasting as long as two hours) might descend are difficult to imagine.

Laboratory studies of captive animals in the 1930s and 1940s

Normally undemonstrative, blue whales sometimes unfurl their flukes upon sounding. (Off western Baja California, 30° 13' N, 116° 24' W, July 9, 1980: M. Scott Sinclair.)

identified many of the adaptations thought to enable marine mammals to make prolonged dives, and supported speculation about others. In the late 1960s a captive bottlenose dolphin (a coastal, shallow-water animal), specially trained for open-ocean release, repeatedly dived on command for periods approaching five minutes and to depths of nearly 300 m, registering the depth of its submergence on a switch at the end of a submarine cable, then returned to its trainer. Studies of this animal, named Tuffy, elucidated many of the anatomical and physiological mechanisms that are thought to make such deep and prolonged dives possible.

The cetacean blowholes have nasal plugs that are closed except when forced open by muscular contractions during respiration. These plugs provide a secure seal during dives. The lungs are elongated and highly elastic. Perhaps surprisingly, they are relatively small in deeper-diving whales, which shows that only a minimal role is played by lung air in prolonged dives. The diaphragm is longer, stronger, and oriented more horizontally than in other mammals, allowing evacuation of a high percentage of lung air in a relatively short time. It has been estimated that from 80 to 90 percent of the lung's air is replaced during each respiration.

In general, cetaceans have a greater ability to transport oxygen across the lung membranes, a higher percentage of oxygen-carrying cells in the blood, and a higher CO_2 tolerance in the cells than have nondiving mammals. During prolonged dives, when oxygen conservation is particularly important, the heart rate and peripheral blood flow are believed to be markedly reduced. This slowing of the heart rate, called bradycardia, has been demonstrated most frequently and dramatically with restrained captive animals forced to dive experimentally; so some sectors of the scientific community suspect the bradycardia may have been induced by trauma, and not be a natural part of the animal's diving reflex. A combination of adjustments limits blood flow to all areas except the heart and brain. On deep dives, when crushing water pressure also becomes a factor, alveoli collapse at about 100 m, preventing absorption of nitrogen into the blood, and making the animal resistant to the bends. Oxygen use is reduced and heat loss inhibited. The degree of development of these characteristics differs from species to species, and is related to their ecological need for deep diving.

THERMOREGULATION: Cetaceans are known to have five mechanisms that regulate their body temperature:

(1) increased insulation (a blubber layer in which blood supplies are minimal, which reduces the chances of heat loss at the surface);

(2) circulatory adjustments to limit heat loss, the most striking of which is a counter-current system in which veins serving the body's periphery are surrounded by arteries, ensuring that heat given up by outflowing blood from the warmer body core is at least partly recovered by nearby inflowing blood;

(3) an increased metabolic rate;

(4) a low ratio of body surface to volume, which is a heat-conserving feature resulting from elimination of hind limbs and reduction of forelimbs; and

(5) decreased respiration rates, so that warm body air is given up less frequently than by terrestrial animals. Together with a variety of other, smaller-scale adjustments, the subtle interplay of these mechanisms allows cetaceans to conserve precious heat.

BIRTH AND CARE OF YOUNG: Thrust as it is from the warmth of the birth canal into a shockingly cold and suffocating medium, the newborn cetacean must be given every possible advantage. Cetacean calves are generally larger, relative to the size of the mother, than the young of terrestrial mammals. Harbor porpoises, for instance, bear calves that are up to 40 percent of the mother's size. This large body size at birth reduces the amount of relative surface area and thereby presumably minimizes heat loss.

Most cetacean calves are known or suspected to be born tail first; so the blowholes are last to emerge, and even in a complicated birth the calf is not likely to inhale too much water. Among at least some of the dolphins and medium-size whales, laboring mothers are closely tended by one or more other adults. When the calf is born, the mother or an attendant may help it to the surface for its first few breaths. Though they may appear somewhat awkward at first, the calves are precocious, immediately capable of swimming and shallow diving.

Cetaceans nurse with a pair of teats concealed in slits along the body wall in the female. The milk, richer in fat than that of terrestrial mammals, is forcibly ejected by strong muscles surrounding

A mother eastern spinner dolphin and her calf. (West of Clipperton Island at 10°25′N, 123°05′W, January 10, 1980: Robert L. Pitman.)

the mammaries. Growth rates of cetacean calves sometimes seem phenomenal. A dolphin may double its length and increase its weight by as much as seven times in the first year. Because of its sheer magnitude, the growth of the blue whale is especially impressive. Newborn are about 7 m long; when weaned at about eight months, they average 15 m, having gained an incredible 90 kg of weight per day.

Calves are closely tended by their mothers and sometimes by other adults, particularly in stressful situations. Sperm and gray whale adults have gained reputations as being aggressive (the gray has been called "devilfish") because of their willingness to defend their young. Such care is part of daily life, smaller calves often hitching rides on the bow waves or in the convection currents created by the stronger adults, a system so efficient that the calf hardly has to move its flukes to keep up with a traveling group. "Babysitting" has been reported, in which an adult other than the calf's mother looks after it while the mother searches for food. Periods of dependence on the parent are long; some odontocetes nurse their young for as long as two years. This protracted dependence ensures a low rate of calf mortality, and compensates for the relatively low reproductive rates and long calving intervals of cetaceans.*

---

* To translate complex biological events into simple numbers requires a thorough understanding of the events and the physiological processes underlying them. Methods most widely used to estimate ages of cetaceans have involved preparation of sliced sections of ear plugs, tympanic bullae, and baleen (for mysticetes), or of teeth and mandibles (for toothed whales), and counting of the number of growth layers they contain. Techniques have continued to be refined, and older methods have often been abandoned in favor of improved methods that reveal more layers. There are problems with the methodology of detecting and counting all the growth layers accurately and with the interpretation of their periodicity. Recently

SENSES: Water does not favor use of some senses that we and other terrestrial mammals take for granted, and in cetaceans some of these are reduced or absent. It appears from their brain structure that odontocetes are unable to smell. Mysticetes, on the other hand, do have some brain structures that can be interpreted as being the nucleus of a lateral olfactory tract, but it is not known whether these are functional. It has been speculated that, as the blowholes migrated to the top of the head, the neural pathways serving the sense of smell may have been nearly all sacrificed.

At least some cetaceans have taste buds, although the nerves serving these receptors have degenerated or are rudimentary. The sense of touch has often been described as weak, but trainers of captive dolphins and small whales often remark on their animals' responsiveness to being touched or rubbed, and both captive and free-ranging cetacean individuals (particularly adults and calves or members of the same subgroup) appear to make frequent contact. This contact may help maintain order within a group, and stroking and touching are part of the courtship ritual in most species. The area around the blowhole(s) is also particularly sensitive, and captive animals often object strongly to being touched there.

Hairs or whisker pits found along the top of the head and sometimes on the chin appear to be tactile organs. They may help the animals sense the air-water interface during the critical moments when they are at the surface for respiration. The tactile whiskers of some river dolphins may augment their sophisticated echolocation system (the use of echoes from sounds of their own making for orientation, navigation, and food finding) by helping them "feel" for food while rooting in bottom sediments.

Vision is developed to different degrees. Mysticetes studied at close quarters underwater—specifically, a gray whale calf in captivity for a year, and free-ranging right whales and humpback whales studied and filmed off Argentina and Hawaii—have obviously tracked objects with vision underwater, and they apparently can see moderately well both in water and in air. However, the position of the eyes so restricts the field of vision in sperm and baleen whales that they probably do not have stereoscopic vision.

The only dolphin species whose vision has been tested experimentally, the bottlenose dolphin, has extremely keen vision in water. Judging from the way it watches and tracks airborne flying fish, it apparently can see fairly well through the air-water interface

developed techniques suggest that many previous readings were in error. Similarly, methods used to estimate reproductive status of cetaceans have involved examination of testes and ovaries for evidence of spermatogenesis or ovulations and pregnancies, respectively. For females some troubling uncertainties persist, particularly about interpreting scars as evidence of ovulation or pregnancy; so reported life spans, ages at sexual maturity, and other statements about a species' life history should be viewed with a measure of skepticism.

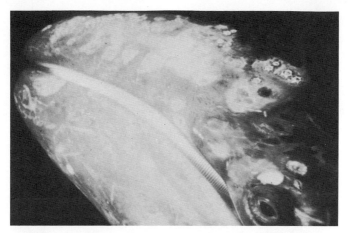

An underwater photographer catches the eye of a young gray whale. Note the orderly row of white baleen. (Off San Diego, California, January 1976: Howard Hall.)

as well. Though preliminary experimental evidence suggests that their in-air vision is poor, the accuracy with which dolphins leap high to take small fish out of a trainer's hand indicates that their vision in air is probably better than one might expect.

The position of the eyes in most dolphins and porpoises, and in some larger toothed whales, suggests that they have stereoscopic vision forward and downward. Eye position in the freshwater dolphins, which often swim on their side or upside down while feeding, suggests that what vision they have is stereoscopic forward and upward. Vision is no doubt more useful to species inhabiting clear pelagic waters than to those living in turbid rivers and flooded plains. The South American boutu and the Chinese beiji, for instance, appear to have very limited vision, and the Indus and Ganges susus are all but blind, their eyes reduced to slits that probably allow them to sense only the direction and intensity of light.

The deterioration of taste and smell, and the uncertain vision in water, is more than compensated for by cetaceans' well-developed acoustic sense. Most species are vocal. Large baleen whales use primarily the lower frequencies, rarely above 5 kHz, and are often limited in their repertoire. Notable exceptions are the nearly songlike choruses of bowhead whales in summer, and the complex, haunting utterances of the humpback whales, much heard of late on record albums and in films. Though the sperm whale apparently produces a monotonous series of high-energy clicks and little else, odontocetes in general employ more of the frequency spectrum and produce a wider variety of sounds than mysticetes. Some of the more complicated sounds are clearly communicative, although what role they may play in the social life and "culture" of cetaceans

From breeding and calving grounds in quiet tropical waters, some humpbacks travel in summer to world's end, along the edges of polar ice. (Hope Bay, Antarctic Sound: Jim Holbrook, courtesy of Hubbs-Sea World Research Institute.)

has been the subject of more wild speculation than solid science.

Mysticetes are not known to echolocate, though some clicklike sounds, apparently made by minke and gray whales, have been recorded. Echolocation has been assumed for all toothed whales, because of detailed experiments with a few species. The nature and sophistication of each species' acoustic capabilities seem related to the demands of its habitat. The different types can be classified as riverine, estuarine, coastal, and oceanic. The most adept echolocator studied to date is the arctic white whale or beluga (sometimes called "sea canary"), which produces sounds exceeding 300 kHz and can resolve extremely fine detail. Nature rarely provides clear evidence of cause and effect, but consider the beluga's need for a versatile navigation system, one that is useful in dark holes and leads in hard ice, in slush zones, in deep open water, in shallow, turbid coastal marine waters, and even in freshwater rivers. Humans attempting to replicate biosonar systems have been humbled by even the simplest systems, the intricate details of which remain unfathomable. Cetacean sounds and their roles in the animals' lives will continue to inspire creative speculation and useful research. Certainly they are as eerie a reminder of the mystery of the sea as can be found.

MIGRATION AND DISTRIBUTION: Cetaceans are among the most mobile creatures in the world, exceeded among large vertebrates only by certain far-ranging birds and fish, and, of course, by technology-aided humans. Some rorquals reportedly can swim as fast as 48 km per hour, at least for short bursts; and the fastest of the dolphins can race along at 28 to 32 km per hour. Their relatively high metabolic rates and carnivorous diets require many cetaceans to range across broad expanses in pursuit of food.

The mobility of most of the so-called great whales—blue, fin, sei, Bryde's, humpback, gray, bowhead, right, and sperm—has been documented by the commercial whaling industry. It has been said that until quite recently the only well-known aspect of the life history of the great whales was where and when they could be found and killed. Although seasonal shifts in distribution and abundance of whales were evident even to the early whalers, it was not until the 1920s that movements of individual animals, bearing numbered darts applied at known locations and times, could be traced. In general, populations of large whales have been found to shift poleward in summer and toward the tropics in winter. Most of this movement is presumably prompted by the availability of generous food supplies in polar seas in summer, and by the relative warmth and calmness of tropical waters in winter. The life cycle of most of these large species seems to dictate a peak period of birth and mating in winter, with summer devoted to a single-minded effort to rebuild their fat stores. Gray whales and humpback whales parade past coastlines (sometimes bringing them near heavily populated urban centers), and their movements constitute a migration in the precise biological sense. On schedule, year after year, their orderly procession across tens of degrees of latitude continues. Other species are much less predictable. The temperate-region right whale, for example, seems to undertake a less well understood seasonal relocation, avoiding extremes of heat and cold, and re-

The gray whale's long and predictable migration makes whalewatching a profitable aspect of tourism on the west coast of North America. (San Ignacio Lagoon, Baja California, February 1980: Steven L. Swartz.)

maining throughout much of the year in regions that provide an adequate supply of food. Sei whales reportedly act like opportunists, varying their migratory schedule and destination from one year to the next, showing up or failing to appear, with little reverence for static laws imposed by human hunters and observers.

Pronounced migrations are not known for most species of smaller whales and dolphins. Changes in their numbers that occur seasonally in many areas seem to result from shifts in their centers of abundance, probably sometimes in response to seasonal changes in water temperature that affect food availability. Coastward movements in many areas by pilot whales, for example, have been closely linked with coastward migrations of spawning squid.

Just as seasonal movements can be partly explained in terms of changes in water temperature and productivity of prey species, so also can some longer-term shifts. Risso's dolphins, for example, reportedly were abundant in the Monterey, California, area at the turn of the twentieth century, absent for much of the next seventy years, then abundant again in the early 1970s. In examining the records of water temperature and fish catches for that period, the late Carl Hubbs concluded that these dolphins were merely the most visible indicator of two periods of exceptionally warm water for the area. Similar long-term changes in the dynamic boundaries of cetacean distribution will probably be discovered.*

Although the movements of most oceanic cetaceans seem related primarily to the availability of food and of suitable calving/nursing sites, some species must respond to additional pressures. The three arctic cetaceans—bowheads, narwhals, and white whales—have learned to live in close proximity to the ice, synchronizing their movements with its formation, disintegration, and drift. Some of the river dolphins—the beiji, tucuxi, boutu, and susu—take advantage of rainy-season inundation to penetrate jungles and floodplains; during the dry season they concentrate in main channels to avoid stranding.

---

*In writing this book, we have come to view numbers—tooth and baleen counts, size ranges, estimates of population, and the like—as necessary evils. It would be unfair not to report such figures where we have them; yet there is an oft-undeserved certainty and finality in numbers committed to print. For some species, particularly the harvested ones that have been extensively sampled, our descriptions of the animals should be relatively close to the truth. But for many species, our statements are based on little more than a few scattered specimens fortuitously caught, or cast ashore, and subjected to some degree of scientific scrutiny. The ranges we give for body size, and tooth and baleen counts, are at least based upon a real sample, however small. Estimates of population size, on the other hand, are frequently figments of the mathematician's imagination, based on very little empirical data. Our own experience in gathering grist for various statisticians' mills, and occasionally in "number-crunching" our own field data, has reinforced our innate skepticism about human ability to census these elusive creatures in most of the areas where they normally live.

Boutús, shown here on a riverbank, inhabit the Amazon River and adjacent inundated rain forests. (Central Amazon, near Manaus, Brazil: Robin C. Best.)

The references to zones of distribution of each species given in the discussions in this book refer to the map preceding p. 1.

STRANDING: The phenomenon called "stranding" is, so far as we know, as ancient as the order Cetacea. Historically, coastal natives apparently looked forward to the casting ashore of whales and dolphins; the ownership of such "royal fish" was an issue of some import in Europe and colonial America. Modern news media have called strandings, especially mass strandings, to the public's attention, and have inspired fanciful speculation about the causes of whale beaching. The truth is that we don't know why an entire herd of whales, usually small or medium-size odontocetes (e.g., pilot whales, false killer whales, melon-headed whales), but occasionally sperm whales as well, will sometimes approach the surf zone as if possessed by an urge to return to the land. It has been argued from records of strandings in the northwestern Atlantic that strandings of a given species become more frequent as that species' local abundance increases. Even when towed back to sea, some stranded individuals refuse to swim to freedom. However, sometimes the stranded animals seem simply to have miscalculated and ventured into a cove or inlet while following prey or avoiding predators.

Scientists wonder whether some strandings may be caused by parasite infestations that interfere with the animals' biosonar navigational system or their sense of equilibrium. Large areas of brain damage due to parasitic flukes have been found. James Mead, who manages the Smithsonian Institution's Marine Mammal Salvage Program, is one of many cetologists who salvage cetacean carcasses from beaches and investigate possible explanations of how they got there. So far, a good general answer has not been found.

CONSERVATION: A century or two ago parts of the marine world were swarming with whales and their smaller relatives, but the thoughtless, short-sighted greed of humans has made the surface of the sea a much duller place today. The spouting and splashing are more sporadic now, totally absent in places. Whales and dolphins have been extirpated in many areas, reduced to relict levels in others. No species is known to have become extinct in modern times, although at least one major isolated population—the gray

The mass stranding of a pod of sperm whales is a spectacular, if sadly moving, sight to behold. (Florence, Oregon, June 1979: Robert L. Pitman.)

whale in the North Atlantic—has been utterly destroyed. The Indus susu is barely hanging on in its shrunken habitat; the European stock of bowhead whales between East Greenland and Novaya Zemlya is all but gone; and in the North Pacific, right whales, gray whales (Korean stock only), and cochitos may soon disappear. The litany of problems could be made much longer. Few species remain unaffected by human activities. If nothing else, a population's viability depends on the health of an ecosystem, and the threat of pollution in its many forms extends today to all the world's oceans and waterways.

This is not a book on conservation, but there are several comments on the subject that we cannot resist making. First, a world system has been developed for controlling, to a welcome degree, the direct exploitation of large whales. The International Whaling

Fin whales are still hunted in parts of the North Atlantic, principally from shore stations at Iceland and Spain. (Hvalfjordhür, Iceland, July 1978: William E. Evans.)

Commission (IWC) agreed in 1979 to the closure of factory-ship whaling, except for a regulated harvest of minke whales in the Antarctic. This measure has all but ended the disgraceful era during which hundreds of thousands of majestic creatures were turned into lubricants, cosmetics, margarine, and meal, with little regard for conservation. One may hope that IWC regulation has forestalled the imminent extinction of the large rorquals, right whale, gray whale, bowhead whale, and sperm whale. In 1979 the IWC adopted a proposal by the Seychelles to recognize the entire Indian Ocean as a cetacean sanctuary, making the whales and dolphins

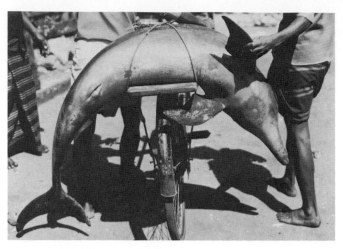

Although common in coastal waters, often near human population centers, bottlenose dolphins are exploited for food in surpisingly few areas. (Beruwala, Sri Lanka, March 3, 1982: Abigail Alling–WWF.)

there safe from commercial hunting by member nations until at least 1989.

Two major challenges remain for the commission: (1) to curb excessive hunting by nonmember nations and "pirate" whalers; and (2) to intervene forcefully, but with sensitivity, in the hunt for bowhead, gray, and humpback whales by native people. The bitter controversy over bowhead whaling in Alaska has focused international attention on the political, cultural, and biological importance of achieving rational regulation of natural-resource use by

A narwhal's long, spiraled tusk, worth several hundred dollars, is sold by a native hunter to a white ivory dealer. Cash income is an important incentive for the hunt. (Admiralty Inlet, eastern Canadian Arctic, August 1975: Randall R. Reeves.)

native peoples. We also hope that the IWC (or its successor) will aggressively extend its purview to smaller cetaceans. Already it has designated the northern bottlenose whale, a species heavily hunted by Norway since the late nineteenth century, a protected stock; and the IWC's panel of scientific experts urged at recent meetings that management of the narwhal and white whale be added to the commission's responsibilities.

Even if the impulse to overexploit marine resources can be subdued, we are left with a world profoundly and permanently changed (some might say scarred) by the past actions of whalers. Trophic relationships among the large whales (to say nothing of seals, fish, and birds) are complex enough to ensure that perturbations caused by the reduction of one species will be felt in many parts of the system. The battering of right, humpback, blue, and fin whales in the southern oceans, for instance, appears to have provided an opportunity for smaller baleen species (sei and minke whales, in particular) and perhaps other animals, such as crabeater seals, to thrive and increase as never before. "Niches" must now be redefined and reallocated as a dramatically disturbed ecosystem seeks a new equilibrium. Things are out of kilter in the oceans now, and will be so for a long time.

Perhaps to balance the brutality with which humans have historically treated cetaceans, many recent commentators have emphasized the sentient gentleness of the whales and dolphins, their supportiveness toward one another, and their supposed lack of aggressiveness. There is evidence that they have all these qualities, and

As we affect the oceans, particularly the yield of food they can sustain, we grow less tolerant of competition from fellow predators. Here a herd of bottlenose dolphins lies slaughtered by fishermen who claimed their own catches were being reduced because of the dolphins' depredations. (Iki Island, Japan, February 1980: Howard Hall, Living Ocean Foundation.)

This harbor porpoise was trapped in a herring weir, but instead of being shot and discarded or eaten, it was tagged and released. (Passamaquoddy Bay, eastern Canada, Summer 1980: Randi Olsen.)

it is a good thing in itself for us to recognize the cetaceans as being fellow creatures, not mere livestock to be slaughtered and consumed. But like other wild creatures, the cetaceans do not always conform to our idealized, often sentimental, image of innocent cuddliness. Scars on the bodies of many toothed whales attest to frequent combat, or at least roughhousing, among males. Humpback and right whales, often portrayed as gentle giants, can be boisterously aggressive, slamming heads and bodies against one another, scraping skin, and even drawing blood. Several species, most notably the killer whale, prey on warm-blooded animals, including other cetaceans. Ascribing cruelty to killer whales for wounding, hounding, and even toying with their quarry is merely reverse sentimentality. Indeed, cetaceans are, like other wild animals, neither good nor bad, and we must resist the temptation to draw them into the ever-changing nexus of judgments we seem determined to impose on members of the human race.

Direct capture of whales may be on the decline, but the incidental or accidental killing of cetaceans continues on a grand scale. The most conspicuous problem of this nature is in the eastern tropical Pacific, where a heavily capitalized multinational purse-seine fishery for tuna intentionally nets, and in the process accidentally kills, tens of thousands of oceanic dolphins each year. Although a sizeable investment has been made, particularly by the United States, in development of alternative, less destructive fishing methods, the tuna-porpoise problem is alive and dangerous in what is now its second decade. Large numbers of dolphins and porpoises (especially phocoenids) are killed inadvertently by foul-

This 4.5-m minke whale, probably recently weaned, became trapped in a near-shore herring weir and was shot. (Blacks Harbour, New Brunswick, eastern Canada, July 29, 1981: Randall R. Reeves.)

ing in gill nets and fish traps. This kind of mortality already may have doomed the cochito, a small porpoise in the Gulf of California, and it threatens to extirpate local populations of several other species in both hemispheres. Antishark nets set off South Africa and Australia sometimes kill large numbers of dolphins. Even the baleen whales sometimes run afoul of fishing gear, most notably off southeastern Canada, where humpback whales, minke whales, and occasionally a right whale or a fin whale become trapped in traps, weirs, or nets. Unlike direct capture, these situations in which cetaceans are killed unintentionally are extremely difficult to control. For some, technological solutions will be found, but others will persist as long as cetaceans and people continue to "fish" the same waters.

The aspect of conservation that drives us closest to despair is pollution. Food-chain magnification, by which even low-level contamination of the smaller prey species becomes concentrated in the tissues of the larger predators, climaxes in marine mammals. Planktonic organisms are carried great distances by winds and tidal currents, and nektonic animals range over great expanses on their own power. Both carry pesticides, heavy metals, and perhaps sometimes radiation and disease-causing organisms to all corners of the globe. Contamination levels in some toothed whales are known to be high; we have no idea at all what this means to their fitness as individuals or their ability to reproduce successfully. Tar balls and oil slicks are now common in the seascape. Cetaceans do not, as we long believed, merely avoid oil by swimming around or away from it. Some whales and dolphins in the western North Atlantic have been seen surfacing repeatedly through oil slicks caused by spillage, making no obvious effort to leave or avoid the contaminated area. On the other hand, gray whales studied off southern California changed their migratory path and their swimming and diving behavior when they encountered patches of oil formed from natural seepage. Whatever direct effects oil may or may not have on cetaceans themselves, its effect on life at any tier of the nutritional pyramid on which cetaceans depend will reduce the environment's carrying capacity by just that much.

The seas are by no means dead, but they are unquestionably less alive than they were when humanity discovered them. We hope that future generations who read this book will know its subjects first-hand, as we have, and that their appreciation for the "spouting fish" will form the basis for an unwavering commitment to their preservation.

Two gray whales encounter a natural oil seep. The effects of oil on whales remain largely unknown. (Off Santa Barbara, California, January 1981: Brent S. Stewart.)

THE BALEEN WHALES

Suborder: Mysticeti

1. The Right Whales

Family: Balaenidae

Bowhead

Balaena mysticetus
Linnaeus, 1758*
Derivation: from the Latin
balaena for "whale"; from the
Greek *mystakos* for "moustache,"
and the Latin *cetus* for "whale."

ZONES 2, 7, AND 8

DISTINCTIVE FEATURES Body rotund; no dorsal fin; in profile, deep depression divides triangular head from rounded back; broad, strongly bowed lower jaws; narrow, arched rostrum; white vest on chin, often with string of black spots; only whale with baleen longer than 2.8 m; blow V-shaped; distribution circumpolar in the Arctic.

DESCRIPTION: This robust whale has been known to reach lengths of 19.8 m, though few exceed 18.5 m. Females are larger than males.

Large individuals can weigh 75 to 100 metric tons. Length at birth is about 3 to 4.5 m.

The enormous head (which occupies a full third of the body length in adults) is dominated by a capacious mouth, housing the longest baleen of any whale; some measured strips were longer than 4 m. The lower jaw is bowed considerably. The lower lips extend beyond and cup the end of the arched rostrum; on the sides the lips rise above and enfold the rostrum. The rostrum is considerably narrower than in the balaenopterids. The blowholes are widely separated, and on windless days the spout is easily recognized as

The bowhead whale's chin is often as white as the ice to which the animal's life history is so closely attuned. (Northern Bering Sea, May 1981: Donald K. Ljungblad.)

V-shaped.

When viewed from the side, the profile of some bowheads consists of two contours separated by a depression: the first, triangular, extends from the tip of the snout to just behind the blowholes; the second, rounded, encompasses the entire back, beginning just behind the head and extending all the way to the flukes. Such a profile may be evident only in adults. Younger animals are often stubbier and in profile have a single smooth contour to the back.

As a rule, the skin is smooth, with no barnacles or encrustations. The throat, chest, and belly lack any sign of ventral grooves; the back lacks any sign of a dorsal fin or ridge.

Bowheads are normally completely black except for a "vest" of

---

* The presence or absence of parentheses around the name of the author who first described the species is not a typographical error. It is biological convention conveying a precise and necessary meaning. The name in parentheses indicates that, although the species name has remained the same since the date of naming, the species has since been assigned to another genus.

uneven white on the chin. Within this vest, near the limits of the white zone, there is often a series of grayish-black to black spots, which on some animals have been likened to a string of beads. Bowheads with lighter gray areas and all-brownish individuals have been reported. In some bowheads there is a light gray band around the peduncle, near the insertion of the flukes.

The baleen is long (4 m or more) and narrow (less than 36 cm across). It is dark gray or black, with long, fine black or gray fringes, in a density of 35 to 70 bristles per square centimeter. Some plates are whitish on the anterior margin, and exhibit a green iridescence in sunlight. There are 230 to 360 plates in each of two rows.

NATURAL HISTORY: Bowheads are not strongly gregarious. They usually travel alone or in groups of no more than about six. Concentrations of fifty or more individuals are sometimes seen on the feeding grounds or when the whales are forced by ice into a restricted area, but such assemblages seem strictly adventitious. What appear to be groups of seven or eight males clustered around an adult female have been observed.

The bowhead's life history is poorly understood. Females probably calve at intervals of at least two years, and most births reportedly occur in spring or early summer.

Although bowheads are migratory, this behavior seems to depend primarily on ice formation and movement. Whaling records suggest that bowheads are segregated to some extent by age and sex during certain phases of their migration. All populations that have been studied winter near the southern limits of pack ice or in polynyas, then press northward in spring as the ice cover breaks up and recedes. In fall they wait as long as possible before they return to their winter haunts, abandoning high-latitude feeding grounds just before the ice forms.

The bowhead is a slow-swimming whale. It reportedly can remain underwater for more than 40 minutes, but is not regarded as a deep diver. Traveling bowheads often raise their flukes on the last dive of a series. These animals have been observed to hang vertically in the water with their heads exposed, and tail-lobbing and flipper-slapping are seen occasionally. Spring vocalizations seem limited to a rather monotonous chorus of low-frequency (less than 1.3 kHz) grunts, moans, and pops; summer and fall underwater sounds are more varied and complex.

Bowheads feed primarily on swarms of small to medium-size zooplankton (euphausiids, amphipods, copepods, mysids, pteropods). Alaskan animals reportedly take about 65 percent euphausiids and 30 percent copepods; so they are in direct competition with abundant seal and arctic cod populations in the area. Although basically skimmers, they do forage in the water column and very near or at the bottom, at least in shallow areas. In such areas, submerged bowheads can often be tracked from their mud trail, and have been seen surfacing at the edge of a large mud boil of their own making with mud and detritus streaming from their mouth. Bottom-feeding bowheads presumably take prey that lives on (rather than in) the sea-bottom mud.

The only natural predator is the killer whale. Starvation from lack of access to feeding grounds and suffocation under ice are probably other natural causes of death.

DISTRIBUTION AND CURRENT STATUS: Bowheads are (or were) circumpolar in the Arctic. Five separate stocks have been proposed: the Sea of Okhotsk; the Bering, Chukchi, and Beaufort Seas or western Arctic; Baffin Bay, Davis Strait, and their adjacent waters; Hudson Bay and Foxe Basin; and the Greenland and Barents Seas. The last of these stocks is very near extinction. The Hudson Bay/Foxe Basin and Sea of Okhotsk stocks may be stable, but the populations are very small (a hundred or less). In Baffin Bay and Davis Strait, there are at least a few hundred left, but their status is uncertain. The only substantial population is that in the Bering, Chukchi, and Beaufort Seas. Whales from this population move northward from the Bering Sea in spring, passing St. Lawrence and Diomede Islands, primarily on their western sides, in three or four pulses or waves. Most turn northeastward in the Chukchi Sea, following the most inshore leads in the ice, round Pt. Barrow in April, and continue along the coastal leads to Banks and Prince Patrick Islands. As ice recedes through summer, they spread south and east, at least into Amundsen Gulf. As ice begins to reform and advance in fall, they move westward, some as far as the northeastern Soviet coast, near Wrangel Island, before yielding to the winter ice by

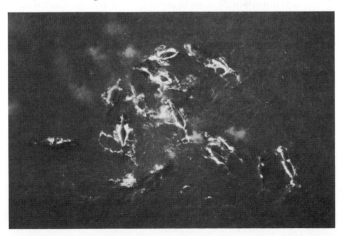

Clouds of whale's breath fill the air above 13 tightly bunched bowheads. (Alaskan Beaufort Sea, 70° 11' N, 144° 37' W, September 14, 1982: Randall R. Reeves.)

moving gradually southward to their winter grounds. This stock, which is estimated to have contained 18,000–36,000 whales at the onset of Yankee whaling in the western Arctic in 1848, is depleted, but may still number between 1,000 and 3,000.

The story of the bowhead's stepwise depletion is well-documented. A European commercial fishery began around Spitsbergen in the seventeenth century, then in Davis Strait during the eighteenth century. By the early nineteenth century, bowhead whaling in those areas had become less profitable, but soon new grounds opened up in northern Baffin Bay and later Hudson Bay. American whalers, who had assumed an active role in the Davis Strait and Hudson Bay fisheries during the late eighteenth and early nineteenth centuries, began bowhead whaling in the Sea of Okhotsk in the mid-nineteenth century and soon thereafter in the Bering Sea and north of Bering Strait. Fortunately, this last episode of irresponsible commercial harvest was ended early in the twentieth century by economic factors: the value of baleen was reduced drastically by the advent of spring steel and plastics, and whale oil became less valuable as petroleum products were developed. Today the bowhead is hunted only from a few coastal villages in northern Alaska, but recent and planned escalation of this fishery poses a further threat to this long-abused species. Sporadic and poorly documented attacks on bowheads by native hunters in the eastern Canadian Arctic, along with potential disturbance by industrial development, make continued survival of eastern Canadian stocks uncertain.

CAN BE CONFUSED WITH: The bowhead's affinity for ice ensures that it will not often be seen near other large whales. Gray whales summer in some of the same areas as bowheads, but the knuckled dorsal ridge and accumulation of ectoparasites on gray whales should make them readily distinguishable from bowheads. The historic summer range of the northern right whale overlapped significantly with the bowhead's winter range, but overlap is highly unlikely in extant stocks. Right whales have callosities on the rostrum and lower jaws, and their baleen is not nearly as long as that of bowheads, although it is longer than that of any other species.

Blue whales are seen occasionally along the ice edge, particularly in the northeastern Atlantic, where bowheads are extremely scarce. Their pale, mottled skin, flat head bisected by a longitudinal ridge, and small dorsal fin, located far back on the body, should make blue whales easy to distinguish from bowheads.

Northern Right Whale

Eubalaena glacialis
(Müller, 1776)
DERIVATION: from the Greek *eu*
for "right" or "true," and the
Latin *balaena* for "whale"; the
Latin *glacialis* for "icy" or
"frozen."

Southern Right Whale

Eubalaena australis
Desmoulins, 1822
DERIVATION: from the Latin
australis for "southern."

ZONES 1, 2, 7, and 8
ZONES 3–5

SOUTHERN RIGHT WHALE

DISTINCTIVE FEATURES: No dorsal fin; large head, strongly arched upper jaws and strongly bowed lower jaws; narrow rostrum; callosities on the rostrum, near the blowholes, and on the chin and lower lips; long, usually dark baleen (up to 2.8 m long); distribution worldwide, in both cold- and warm-temperate regions.

DESCRIPTION: These robust whales can grow to about 17 m long and weigh close to 100 tons. Females are larger than males. Length at birth is 4.5 to 6 m.

The body is rotund, with measured girths reportedly 60 percent or more of body length, but tapers to a relatively narrow tail. The head is large, about one-fourth or more of body length. The rostrum is long, narrow, and highly arched. The strongly bowed lower lips close over (enfold, in a sense) the rostrum on either side. There is a series of growths called callosities, the largest of which is called the "bonnet," on the rostrum in front of the two widely divergent blowholes. Callosities are also present on the chin, on the sides of the head, on the lower lips, above the eyes, and near the blowholes. Their arrangement apparently differs for

The head of a southern right whale. (Near South Georgia, southwestern Atlantic, March 1977: Frank S. Todd, Sea World, Inc.)

each individual. Cyamid crustaceans, called whale lice, frequently live on the callosities, making them appear white, orange, yellowish, or pinkish. Hairs are numerous on the chin and upper jaw. The spout is bushy and V-shaped. It can be 5 m high.

There is no dorsal fin or ridge on the back, nor is there evidence of ventral grooves. The flippers are large and broad, with an angular outer edge. The flukes, which are often lifted as the animal dives, have a broad, smooth, and concave rear margin, with a deep notch between them.

The skin is black to brown, sometimes mottled, with irregular white patches on throat and belly, and less frequently on other parts of the body. Albinism, or at least piebalding in the extreme, has been observed in the South Atlantic.

The baleen is up to 2.8 m long. It ranges in color from dark brownish through dark gray to black; some of the anteriormost plates are sometimes white. When the animal swims, mouth agape near the surface, the baleen sometimes appears pale, even white. There are usually 220 to 260 long, narrow plates, fringed by long, fine, grayish bristles, in each side of the upper jaw.

The nomenclature and systematics of right whales are unresolved. There may be three valid species, or two, or only one. There is no known external morphological basis for recognizing more than one species, but it has been suggested that southern right whales have callosities along the upper surface of the lower lips more consistently than northern right whales.

NATURAL HISTORY: Aggregations of as many as 100 individuals were seen early in this century on feeding grounds, but small groups of two to a dozen are now more common. Long-term social

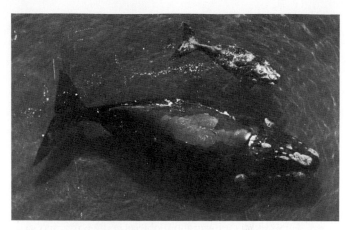

A passive mother right whale dozes at the surface (note the dry back) as her calf swims round her, trailing streams of bubbles from its blowholes. (Off Peninsula Valdez, Argentina: Roger S. Payne.)

relationships, other than a strong female-calf bond, are undocumented. However, distribution tends to be highly concentrated; so there may be a kind of "herd" integrity within the population.

There is little definite information about the reproductive biology of right whales. The slowness of their recovery from over-exploitation suggests that either the reproductive rate is low or the mortality rate is high. Observations off Argentina suggest that the calving interval is usually three years, but it can be two or four years. Calving occurs in winter or spring at relatively low latitudes, and mating may also peak during this season. However, recent observations of intense sexual interaction in the lower Bay of Fundy during August and September suggest that some conceptions may occur in summer. In the well-studied Argentine population, most females return to the breeding area every third year, whereas males usually return annually.

Much of what is known about the behavior of right whales was learned in the southern hemisphere, especially at Peninsula Valdez on the central coast of Argentina, where a large wintering concentration was discovered in 1969. Right whales come to these shallow, sheltered waters in July and stay until November, during which time females with calves remain isolated in what has been described as a "nursery." In a different area, mating is frequently observed, with half a dozen males sometimes courting a single female simultaneously. Feeding also has been reported in the vicinity of Peninsula Valdez. Since 1980, much of the behavior reported for Peninsula Valdez has also been seen in the lower Bay of Fundy in August and September.

There is a seasonal north-south pattern in the movements of right whales but their migrations seem much less regular and coherent than those of humpback, bowhead, and gray whales. Right whales are most likely to be encountered at latitudes lower than 50° during winter, and at latitudes higher than 40° during spring, summer, and fall, but in the middle parts of their range, 40°–50°, they can be encountered in any season. Also, even migrating individuals seem to remain in an area for days or even weeks at a time.

These whales are slow swimmers, and can exceed 12 km/hour only in short bursts. They are not regarded as deep divers, since they find most of their prey not far below the surface, and maximum submergence times are about 20 minutes. They are surprisingly acrobatic and demonstrative for such bulky creatures, often breaching, flipper-slapping, and tail-lobbing. "Head-standing" is a common behavior of southern right whales, and is also observed occasionally in the North Atlantic. In Argentina, right whales are frequently found inside the surf line, though they almost never strand. They exhibit curiosity about objects in the water, and seem to interact peaceably with smaller cetaceans and pinnipeds, which often frolic behind them or catch rides in their bow waves. Vocalizations seem limited to low-frequency moans and belch-like utterances.

Right whales are specialized eaters, taking copepods if available and euphausiids as a second choice. They apparently shun fish and the larger invertebrates altogether. Attacks by killer whales have been reported, but at least some of these are probably unsuccessful.

DISTRIBUTION AND CURRENT STATUS: Right whales were once abundant along all major land masses in the temperate latitudes of both hemispheres. The Atlantic and Pacific populations in the northern hemisphere are geographically separate, and are sometimes considered different subspecies, or even species. In the eastern North Atlantic, they ranged from the Azores, Madeira, and northwestern Africa, occasionally entering the Mediterranean Sea, to the waters around Iceland, the Faeroes, Shetlands, and Hebrides, off southeastern Greenland, and between Spitsbergen and North Cape (Norway). The Bay of Biscay was an important wintering area. Several right whales killed near Madeira and a few sightings between Spain and Norway constitute the only recent evidence of the species' continued existence in the eastern North Atlantic.

In the western North Atlantic, the species ranged (and still does in very small numbers, perhaps several hundred) from Florida and the Gulf of Mexico to the Gulf of St. Lawrence and the coasts of Nova Scotia, Newfoundland, Labrador, and possibly southern Greenland. Cape Cod Bay and waters off Cape Cod are still visited regularly by right whales. At least a few dozen right whales concentrate in the lower Bay of Fundy, near Grand Manan Island, during August and September, and the vicinity of Browns Bank, off the southern tip of Nova Scotia, appears to be another important assembly area. Right whales have been seen recently inside the Gulf

The broad flipper of a northern right whale, on its side near the sea surface. (Lower Bay of Fundy, September 11, 1982: Randi Olsen.)

of St. Lawrence and in Trinity Bay, Newfoundland.

In the eastern North Pacific, right whales were found from central Baja California to the Gulf of Alaska and into the Bering Sea. Japanese and Soviet catcher boats logged more than 100 sightings of right whales in the area west of about longitude 150° W and north of latitude 50° N between 1941 and 1968. Records from east and south of that area are more scarce, consisting since the mid-1950s of a handful of sightings, from the coasts of Washington, Oregon, southern California, and northwestern Baja California. A sighting in Hawaii several years ago is the first for that area in this century.

In the western North Pacific, the southern limit was probably near Taiwan or the Bonin Islands; the northern limit along the Kamchatka peninsula, the Commander Islands, and into the Sea of Okhotsk. Some are still seen off Hokkaido, the Kuriles, and Kamchatka, and in the Sea of Okhotsk. The total remaining North Pacific population has been estimated at an alarmingly low 200 to 250.

The southern right whale apparently is isolated from the northern right whale by the tropical belt between roughly 20° N and 20° S (there are a few historical records of right whales in the tropics, but they are difficult to document). It is circumpolar in the southern hemisphere between 20° S and at least 55° S. Different migratory streams seem to represent isolated stocks. Historically, right whales were found seasonally off South America (Chile, Argentina, Brazil), South Africa (Cape of Good Hope), Australia (southern and western regions, and New South Wales and Tasmania), and New Zealand. They also clustered around temperate and subantarctic islands, such as Tristan da Cunha, South Georgia, Kerguelen, Crozet, Campbell, Auckland, and the Chathams, as well as in such offshore areas as Chatham Rise, between New Zealand and Campbell Island.

The total southern-hemisphere population may have exceeded

100,000 before commercial exploitation on a major scale began in the early nineteenth century. Like the northern-hemisphere stocks, southern populations were very nearly exterminated in spite of their far-flung and rather dense distribution. During the last twenty years a slow recovery has become evident in some areas, with right whales seen in small numbers in many of their old haunts, including Campbell Island (a few score), and the coasts of western Australia, New Zealand, Chile, South Georgia, Tristan da Cunha, and South Africa. A few have also been sighted by pelagic whalers in the South Pacific and Indian Oceans. The present total southern-hemisphere population probably numbers several thousand. Golfo San Jose, enclosed by Peninsula Valdez in Patagonia, is a wildlife sanctuary, and one of the best places in the world to watch right whales; more than 580 individuals have been identified there since 1969.

Space does not allow us to detail the long and depressing history of the right whale's undoing. Its approachability, relative docility, and buoyancy when dead made it the "right" whale for pursuit by coastal and pelagic whalers throughout its range. Most stocks did not even last long enough to experience the deadly efficiency of modern whaling techniques. Since 1937 they have enjoyed almost complete protection, but their failure to rebound more convincingly suggests that it may have come too late.

CAN BE CONFUSED WITH: The right whale, with its finless back and huge head marked by the distinctive callosities, is unlikely to be confused with other species in temperate latitudes. In the northern extremes of its range, however, it might be confused with the bowhead. Chances for confusion should be decreased by the fact that the bowhead has already moved north before the right whale

When seen so clearly, a right whale is difficult to misidentify. (Off Fort Pierce, Florida, February 1967: courtesy of Miami Seaquarium.)

reaches the northern latitudes in which the two species' ranges overlap. When there is doubt, the right whale's callosities, especially the prominent "bonnet," are the best keys for distinguishing it from the bowhead. There is some possibility of confusing right and gray whales, but only in the North Pacific. The knuckled ridge along the spine of the gray whale distinguishes it from the smooth-backed right whale. The humpback's bushy, sometimes V-shaped spout, together with its habit of lifting its flukes when diving, can cause confusion. However, closer inspection should reveal the humpback's knobby skin, dorsal fin or hump, long narrow flippers, and distinctively shaped flukes.

Pygmy Right Whale

Caperea marginata
(Gray, 1846)
DERIVATION: from the Latin *caperea* for "to wrinkle," referring to the texture of the ear bone in this species; from the Latin *marginata* for "enclose with a border," referring to the dark border observed on the baleen of the type specimen.

ZONES 3 TO 5

DISTINCTIVE FEATURES: Bowed lower lips; arched, tapered rostrum; small falcate dorsal fin; the jaw configuration distinguishes it from the balaenopterids, the presence of a dorsal fin from other balaenids; southern hemisphere, temperate distribution.

DESCRIPTION: This whale has features of both rorquals and right whales. Its unique skeletal features have led some investigators to classify it in a family by itself. The largest known female specimen was 6.4 m long; since female baleen whales are larger than males, this may be the approximate maximum for the species. No newborns or even near-term fetuses have been examined; so length at birth is unknown.

The head is not so disproportionately large as it is in other right whales, reaching no more than one-quarter of the total length. The lower jaws are bowed, and the upper jaw is arched, a feature that seems to become more exaggerated with age. The line of the mouth extends to behind and below the eye. The lower jaw projects slightly beyond the upper jaw. In the only living animal viewed underwater, the ventral outline of the throat was concave anteriorly, becoming convex posteriorly. It has been suggested that the two indistinct longitudinal furrows on the throat—caused by three mandibular ridges—might be analogous to the throat grooves of rorquals and gray whales.

A small dorsal fin, with a smooth anterior margin and a falcate posterior margin, is placed well back on the streamlined body. The flippers are small, narrow, and slightly rounded at the tips; the tail flukes are broad and full, with a well-formed median notch.

Coloration is gray or dark gray dorsally, and white ventrally. The lower half of the lower jaw is noticeably lighter than the upper half. When the mouth is open, a white band is discernible just below the line of the upper jaw; this is exposed baleen gum. The flippers are darkly pigmented, and thus well-demarcated from the body at their insertion. Limited observations of animals of different ages suggest that skin color may darken with age.

There are about 230 strips of yellowish white baleen in each upper jaw. The longest plates measure about 68 cm; the shortest, a few centimeters. The baleen has been said to be more flexible, more elastic, and tougher than that of any other whale.

NATURAL HISTORY: Only a few dozen specimens of this species have been examined by scientists, and sightings at sea are extremely rare; so little is known about the natural history of the pygmy right whale.

Groups of as many as eight animals have been seen, but most encounters have been with lone individuals or pairs. Pygmy right whales have been observed in the company of dolphins and pilot whales, and at least once with a female sei whale and calf.

There is no information on the reproduction of pygmy right whales. Their size at weaning has been estimated at 3.2 to 3.8 m.

There is some evidence for an inshore movement in spring and summer, but no long-distance migration has been documented. While inshore, the pygmy right whale seems to have a strong preference for sheltered, shallow bays. It has been found in water temperatures ranging from 5° to 20° C.

Its behavior has been described as "unspectacular," and perhaps this is why it is so rarely observed. The small blow is often inconspicuous. Though the pygmy right whale does not seem to be a deep or prolonged diver, it apparently spends little time at the surface. The dorsal fin is often not seen at all when the animals surface for air. Occasionally upon surfacing this whale "throws" its snout clear of the water and thus exposes the end of the chin. At such times a flash of white, attributed to the tongue or the lining of the mouth, can be seen.

Swimming behavior was described from underwater observations of a temporarily restrained animal. As it swam it exhibited an unexpected flexing of the entire body, rather than just the tail region, with waves of motion passing from the head to the flukes, increasing in amplitude as they did so. The flexing made the whale's movements seem jerkier and less graceful than those of other observed baleen whales. Of course, this behavior may have been artifactual (due to confinement) or unrepresentative (i.e., peculiar to this individual). Swimming speed for this species seems to range between 5 and 8 km/hour, though the observed individual seemed capable of extremely rapid acceleration.

The only known food is copepods. Natural mortality factors are completely unknown.

DISTRIBUTION AND CURRENT STATUS: The pygmy right whale is known only in temperate waters of the southern hemisphere. It does not appear to venture south of the Antarctic Convergence (the dynamic boundary between Antarctic and the sub-Antarctic water masses, usually at about 60° S) or north of the 20° C isotherm (roughly 30° S). Its presence around Tasmania has been noted in all seasons, and it is known to be present at least seasonally along the coasts of South Australia, New Zealand, and South Africa (between False Bay and Algoa Bay). There are pelagic records of the pygmy right whale from the South Atlantic and southern Indian Oceans.

The pygmy right whale has never been and is not currently exploited. Some incidental mortality apparently results from coastal netting operations in South Africa.

CAN BE CONFUSED WITH: Many sightings of this whale at sea probably have been incorrectly logged as minke whales. If all that is seen is the blowhole, back, and dorsal fin, there is little hope of differentiating between the two species. When present, the minke whale's white flipper band (not all minkes in the southern hemisphere have it) is diagnostic, as is the jaw configuration in the pygmy right whale. Unfortunately, these key field marks may not always be readily observable.

2. The Rorqual Whales

Family: Balaenopteridae

Blue Whale

Balaenoptera musculus
(Linnaeus, 1758)
DERIVATION: from the Latin
balaena for "whale," and the
Greek *pteron* for "wing" or "fin";
from the Latin *musculus*,
diminutive of *mus* for "mouse"

(perhaps meant as a joke, since
the blue whale is so unlike a
mouse, but also sometimes
taken to mean muscular).

ALL ZONES

DISTINCTIVE FEATURES: Body huge, mottled bluish gray; head flat in front of blowholes, viewed from side; head broad and nearly U-shaped, viewed from above; dorsal fin a small nubbin located in the last fourth of back.

DESCRIPTION: These are the largest living animals. Before stocks were severely depleted by whaling operations, antarctic blue whales reached 30.5 m in length, with estimated weights of over 160 tons. Those in the northern hemisphere are smaller, reaching 24 to 26 m in length. In all known populations, females are slightly larger than males of the same age. Both sexes are thought to become sexually mature at about ten years of age, at which time the males average 22.5 m and the females 24 m. Newborn average 7 m.

Viewed from above, the blue whale's rostrum is broad, flat, and nearly U-shaped, resembling a Gothic arch that is slightly flattened on the tip, with a single ridge extending from the raised area in front of the blowholes toward, but not quite reaching, the tip of the snout.

The body is very broad in dorsal aspect, and is dominated by two slightly concave zones above the lungs, which appear to contract and expand as the animal respires. These zones accentuate the spine. The tail stock is narrow in dorsal aspect, the flippers long and slim, and the flukes broad and triangular, with a slightly notched but otherwise smooth rear margin. The dorsal fin is extremely small (usually less than 33 cm high), and may range in shape from nearly triangular to moderately falcate. It is always located so far back on the animal's tail stock that it is seldom visible until the animal is about to begin a dive. There are 55 to 68 ventral grooves extending at least to the navel.

Blue whales are a light bluish gray overall, mottled with gray or grayish white. When viewed dorsally, coloration on the back of at least some animals appears to include, on the shoulders just behind the blowholes, a dark "W" oriented anteriorly. Some animals may have yellowish or mustard coloration, primarily on the belly, caused by diatom accumulations formed during long stays in cooler waters. The undersides of the flippers are light grayish blue to white. The relatively short, stiff, and coarsely fringed baleen plates are all black. There are 260 to 400 plates of baleen per row.

Portions of southern-hemisphere populations of blue whales have been described as a separate subspecies, the pygmy blue whale (*B. m. brevicauda*), because of their larger baleen plates, shorter tail

A blue whale fills its lungs before another dive. (Off Bahia San Quentin, Baja California, June 3, 1965: Kenneth C. Balcomb.)

An encounter between superlatives: a great blue whale, the largest of animals, under attack by a pod of killer whales, the largest of predators. (Near Cabo San Lucas, Baja California, May 1978: courtesy of Hubbs-Sea World Research Institute.)

region and longer trunk, extra caudal vertebra, and different growth and maturation rates.

NATURAL HISTORY: Blue whales usually occur singly or in pairs, though several individuals or pairs may be found within a few kilometers of one another in rich feeding areas.

Females give birth to a single calf every two to three years. Gestation lasts for approximately twelve months. The calf is closely tended until fully weaned some eight months after birth, at an average length of 15 m. During that period it will have gained as much as 90 kg per day.

Their enormous energy requirements force most blue whales to migrate long distances, which many do on predictable schedules and along well-known routes between rather restricted high-latitude summering grounds and low-latitude wintering grounds.

The blue whale's blow can be as much as 9 m high, slender, and vertical. It is not bushy like those of humpback, gray, and right whales. The blowing and diving patterns of blue whales vary with the speed of movement and the activity of the whale. If the animal is moving slowly, the blowholes and part of the head may still be visible when the dorsal fin breaks the surface, and the animal may settle quietly into the water without exposing the last portion of the tail stock or the flukes. If the animal is moving more quickly or is about to begin a long dive, the blowholes disappear and the dorsal fin emerges briefly just before the animal lifts its tail stock and flukes slightly above the surface before slipping out of sight. Blue whales lift the flukes differently from the humpback, right, or sperm whale: the latter species raise the flukes high out of the water, and usually descend at a steep angle as they begin a long dive; blue whales lift the flukes only slightly.

Blue whales are relatively shallow feeders, preying almost exclusively on krill, small shrimplike crustaceans, most of which are distributed within 100 m of the surface. A single blue whale might consume up to eight tons of krill in a single day.

Remarkable as it may seem, these magnificent giants are sometimes attacked and killed by killer whales. In one dramatic attack off Baja California, nearly forty killer whales prevented a large blue whale from diving for more than an hour while they bit at its flukes and lips.

DISTRIBUTION AND CURRENT STATUS: Blue whales occur in all oceans, primarily along the edge of continental shelves and along ice fronts, but also venturing into deep oceanic zones and shallow inshore regions. Three major populations, each composed of several stocks, are recognized: North Pacific, North Atlantic, and southern hemisphere.

In the North Pacific, the bulk of the population appears to summer from central California through the Gulf of Alaska to the Aleutians, with some individuals venturing as far south as 33° N, near the California Channel Islands on the east, and others as far north as St. Lawrence Island in the Bering Sea. This population winters from the open waters of the midtemperate Pacific south at least to 20° N. Another group may be resident between 10° N and the equator.

In the North Atlantic, blue whales have been reported from San Cristobal in Panama and the Cape Verde Islands to the pack ice, though they normally are more restricted in their range. They summer from at least the Gulf of St. Lawrence and southern Greenland north to the edges of pack ice in the west and from Iceland, the British Isles, and southern Norway north to Spitsbergen and the Murmansk coast of Russia in the east. Large numbers were killed in waters off northern Norway, Iceland, and the British Isles during the late nineteenth and early twentieth centuries, primarily during summer months. Though unknown, their winter range in the North Atlantic is assumed to extend from midtemperate latitudes, perhaps south into the tropics. There are no definite records of sightings from the coast of North America south of New Jersey or from the West Indies, although there are several probable records of stranded blue whales for the north coast of the Gulf of Mexico.

In the southern hemisphere, blue whales generally stay south of 40° S during summer, and move northward ahead of advancing pack ice in winter, extending their range to Rio Grande do Sul in Brazil, Ecuador, and South Africa, and less commonly to near Australia and New Zealand.

Sightings or strandings have been recorded in the northern Indian Ocean, including the Arabian Sea and Bay of Bengal, and blue whales have been seen just south of Madagascar in summer. Pygmy blue whales are found in the southern Indian Ocean, and seem to inhabit at least the waters between 30° E and 80° E and north of 52° S.

Despite a reduction of hunting in the southern hemisphere beginning more than 25 years ago, and complete protection of all populations by international agreement in 1965, blue-whale stocks are still depressed. Rough estimates suggest that the population of more than 5,000 that inhabited the North Pacific before commercial exploitation now numbers 1,200 to 1,700; only a few hundred are thought to survive in the North Atlantic; and of the 200,000 in the southern hemisphere in the nineteenth century, only about 9,000 (half of them pygmy blue whales) remain.

Blue whales are seen with some regularity in at least three areas: deep coastal canyons off central and southern California, far inside the Gulf of St. Lawrence, and on the whaling grounds for fin whales in Denmark Strait. In the Gulf of St. Lawrence they often can be sighted in late summer and early fall from clifftops along the north shore.

CAN BE CONFUSED WITH: At sea, blue whales may be confused with fin whales and sei whales. Adult blue whales should be easy to distinguish by size alone from immature finbacks and from sei whales of any age. Fin whales are an even gray on the back and white on the ventrum, with asymmetrical head coloration; the right lower lip is white, the left gray. Also, they tend to have a sharper, more V-shaped head, and a comparatively prominent dorsal fin. Dead fin whales can be distinguished from blue whales by the gray to white appearance of much of their baleen, in contrast to the solid black baleen of the blue whale.

Large sei whales often are heavily scarred, and can have mottled skin. Their dorsal fin is usually quite tall, however, and is placed much farther forward on the body than the blue whale's. The rostrum is slightly arched rather than flattened. The sei whale's gray baleen is much more supple and finely fringed than that of the blue whale.

Japanese who have whaled for both blue and pygmy blue whales are convinced that the two forms can be distinguished at sea. They report, for example, that the pygmy blue whale's head is shorter and more blunt than that of a blue whale, appearing broad relative to length, and that the pygmy blue whale is marked by more and lighter spots, and has a noticeably shorter tail stock than the blue whale.

Fin Whale

Balaenoptera physalus
(Linnaeus, 1758)
DERIVATION: from the Greek
physalos for "rorqual whale"
or "a kind of toad that puffs
itself up."

ALL ZONES

DISTINCTIVE FEATURES: Right lower lip and palate white; right front 20 to 30 percent of baleen plates yellowish white; head V-shaped when viewed from above; frequently has grayish white chevron on back behind head; upon surfacing usually presents wheel-like silhouette; does not unfurl flukes on last dive.

DESCRIPTION: Fin whales grow to 24 m and 26.8 m in the northern and southern hemispheres, respectively. Females grow slightly larger than males. Both sexes reach sexual maturity between six and twelve years of age, at average lengths of 17.7 m for males and 18.3 m for females. Newborn are about 6 to 6.5 m.

The rostrum is narrower and more V-shaped than that of the blue whale, and it has a prominent median ridge. The top of the head is flat, though slightly less so than the blue whale's, and has none of the downward turn at the tip characteristic of the sei.

The back is distinctly ridged from the dorsal fin to the flukes, prompting the common name "razorback." There are 56 to 100 slim ventral grooves extending at least to the navel.

The dorsal fin may be more than 60 cm tall, usually angled less than 40° on the forward margin. Its usual location is about one-third the body length forward from the fluke notch, and in traveling whales it appears on the surface shortly after the blow.

Fin whales are dark gray to brownish black on the back and sides, with none of the mottling that is present on blue whales, and they are rarely as heavily scarred as sei whales. Along the backs of many individuals, just behind the head, there is a grayish white chevron, with the apex along the midline of the back and the arms of the chevron oriented posteriorly. This chevron may be visible as the animal surfaces. The undersides, including the undersides of the flukes and flippers, are white. On the head the dark coloration is

Fin whales commonly enter embayments and inshore passages; these two show some of the variability in dorsal fin shape among individuals. (In Head Harbour Passage, lower Bay of Fundy, Summer 1981: Randi Olsen.)

markedly asymmetric, reaching farther down on the left than on the right side. The right lower lip, including the mouth cavity, and the right front baleen (approximately 20 to 30 percent) are white or yellowish white. Occasionally the right upper lip is also white, and in some animals a brush of light gray, continuous with the white area of the right side of the head, sweeps onto the dorsal surface of the neck.

The remainder of the baleen plates on the right side, and all those on the left side, are striped with alternate bands of yellowish white and bluish gray. The fringes of the plates are brownish gray to grayish white. The 260 to 480 baleen plates on each side reportedly reach maximum dimensions of 72 cm in length and 30 cm in width.

NATURAL HISTORY: Fin whales are sometimes found singly or in pairs, but more often in pods of three to seven individuals. Several pods, totaling as many as fifty animals, may be concentrated in a small area.

Fin whales calve and mate in winter, mostly in temperate waters. Every second or third year after attaining sexual maturity, females bear a single calf after a twelve-month gestation period. The calf is weaned at between six and eight months of age.

Like other large baleen whales, most fin whales annually move toward the poles in spring and toward the equator in fall. Some northern-hemisphere populations also apparently shift to areas of concentration inshore in winter.

A fin whale's blow is tall and impressive, a jet of vapor 4 to 6 m high that often can be seen at a great distance. Its shape is like an elongated, inverted cone.

Fin whales dive to depths of at least 230 m. In general, they probably dive deeper than either blue or sei whales. However, the surfacing attitude of a given whale of any of these species depends on whether it happens to be traveling, resting, feeding near the surface, or feeding at depth.

When they are moving leisurely at the surface, fin whales expose the dorsal fin shortly after the appearance of the blowholes. When they are surfacing from a deeper dive, however, they emerge at a steeper angle, blow, submerge the blowholes, and then arch the back and dorsal fin high above the surface. Fin whales do not show their flukes when beginning a dive.

Unlike blue or sei whales, fin whales do breach on occasion. When they leap clear of the water, fin whales usually re-enter with a resounding splash, like that made by humpback and right whales, not smoothly and head-first like minke whales.

Fin whales are reportedly one of the fastest of the big whales (sei whales may be slightly faster), possibly reaching burst speeds in excess of 32 km per hour. A radio-tagged fin whale near Iceland traveled at average speeds of 3.6 to 14.6 km per hour, covering up to 292 km in one day. Like the other large rorquals, the fin whale did not become an important commercial species until the comparatively recent development of fast catcher boats.

Fin whales eat a wide variety of foods, including krill and other

Dolphins, like this white-beaked dolphin, associate with and sometimes hitch rides on the bow waves of great whales, such as this fin whale. (Near Blandford, Nova Scotia, August 1976: Kenneth C. Balcomb.)

invertebrates, capelin, sand lance, squid, herring, and lanternfish. They are themselves victims of occasional attacks by killer whales.

DISTRIBUTION AND CURRENT STATUS: Fin whales have a world-wide distribution, though they tend to be less common in tropical than in temperate, arctic, and antarctic waters.

In the eastern North Pacific they winter from at least central California southward to 20° N, and summer from central Baja California into the Chukchi Sea, with concentrations off central California, in the Gulf of Alaska, in Prince William Sound, and along the Aleutian Islands. One population in the Gulf of California is considered resident.

In the western North Pacific they winter from the Philippine Sea to at least 40° N, including concentrations in the East China Sea, the Yellow Sea, and the Sea of Japan, and along the Pacific coast of Japan; they summer from about 35° N to the Chukchi Sea, including the Sea of Okhotsk and the vicinity of the western Aleutians.

In the western North Atlantic, they winter from the ice edge south to Florida and the Greater Antilles, and into the Gulf of Mexico, primarily in offshore waters. They summer from below the latitude of Cape Cod to the Arctic Circle, and appear to be concentrated between shore and about the 1,800 m curve from at least 41°20′ N to 57°00′ N. There are sizeable concentrations off the northeastern United States, eastern Canada, and western and southeastern Greenland.

They are present in the Mediterranean Sea and in the eastern North Atlantic from the latitude of the Strait of Gibraltar to southwestern Norway in winter. Although fin whales are present in the Mediterranean year-round, they apparently migrate to more northerly waters along the eastern European coasts. There are concentrations in summer off Iceland, western Scandinavia, Jan Mayen, and the Spitsbergen archipelago and in the northwestern Barents Sea.

In the southern hemisphere fin whales migrate from summering grounds in the Antarctic (mostly between 47° S and 60° S) past New Zealand into the southwestern Pacific; into the central South Pacific; along South America as far as Peru on the west coast and Brazil on the east coast; to the central Atlantic off the west coast of South Africa; and into poorly known regions of the southern Indian Ocean above Kerguelen and Heard Islands, where the majority winter between 20° S and 40° S.

After the severe depletion of stocks of blue whales, particularly in the Antarctic, fin whales became the most commercially valuable baleen whale, and long constituted a major portion of the world's whaling catch. Their populations were severely reduced in most areas, particularly in the southern oceans, and some measure of protection has been extended to those that remain. Large-scale whaling for the species has stopped, though hunting for fin whales continues at a modest, stable level from a shore station at Hvalfördhur, Iceland. Conservation problems persist because of continued whaling by Spain in the eastern North Atlantic, and by Korea and perhaps the People's Republic of China in the western North Pacific. There are no active major fisheries in the southern hemisphere except those conducted by pirate whalers, whose activities are unregulated and largely unreported.

CAN BE CONFUSED WITH: Fin whales may be confused with blue whales, sei whales, and, in the more tropical portions of their range, Bryde's whales. Blue whales have lighter, mottled blue-gray body coloration, lacking a chevron; an all-dark and broadly U-shaped head; and a tiny dorsal fin, which in traveling whales appears late at the surface after the blow.

Close examination allows one to distinguish fin whales from most Bryde's whales, which have three ridges on the rostrum. They may usually be distinguished from sei whales by the following characteristics: the fin whale has a white right lower lip and mixed-colored baleen (white in right front, streaked with yellowish white and bluish gray elsewhere), but the sei whale has gray lips and uniformly ash-black baleen with fine gray bristles; the fin whale has 56–100 ventral grooves, which extend to or beyond the navel, but the sei has only 32–60, which end anterior to the navel. However, there are intergrades of these characteristics; so one alone often is insufficient for identification.

At sea subtle differences in appearance of the dorsal fin and in surfacing and diving behavior during traveling and feeding may combine to help one distinguish fin and sei whales. Fin whales usually rise obliquely; so the top of the head breaks the surface first. After blowing, the animal arches its back and rolls forward, exposing the slightly falcate dorsal fin, which usually forms an angle of less than 40° with the back and is about one-third of the body length forward from the fluke notch. Although erratic, fin whales usually dive for five to fifteen minutes, blow three to seven times at intervals of up to several minutes, and then resubmerge.

Primarily skimmer feeders, sei whales usually rise to surface at a shallow angle; so the dorsal fin and head become visible almost simultaneously. The fin usually forms an angle with the back that is greater than 40° and is located just more than one-third of the body length forward from the fluke notch, and the head is slightly downturned at the snout. Even when starting a long dive, a sei whale does not usually arch its back as much as a fin whale. A feeding sei whale's dives, of three to ten minutes each, are usually separated by a long series of evenly spaced respirations. During such periods the whale is often visible just below the surface, even when it is making longer dives.

Sei Whale

Balaenoptera borealis
Lesson, 1828
DERIVATION: from the Latin
borealis for "northern."

ALL ZONES

DISTINCTIVE FEATURES: Body dark gray on back, often with ovoid grayish white scars; dorsal fin prominent, almost one-third of body length forward from the fluke notch, usually forming an angle of more than 40° with the back, often visible as the whale blows: frequently leaves long series of "tracks" on surface during shallow feeding dives; tip of snout slightly turned downward, in lateral profile, with single rostral ridge.

DESCRIPTION: Sei (pronounced "say," after *seje,* the Norwegian word for "pollack," with which they are often found off northern Norway) whales reach 17.1 m (males) and 18.6 m (females) in the northern hemisphere, and 17.7 m (males) and 21 m (females) in the Antarctic. Males and females become sexually mature at six to twelve years of age, by which time they are 12 to 13.1 m or 12.7 to 13.7 m long, respectively. Newborn are 4.5 to 4.8 m.

In dorsal view the snout is less rounded than that of the blue whale, and perhaps less acutely pointed than that of the fin whale. When viewed from the side, it appears slightly arched, with the downward turn accentuated on the tip. There is a single prominent

Head-on view of a sei whale in the Antarctic. (Asahi Shinbun Press; courtesy of H. Omura.)

rostral ridge from the blowholes to the tip of the snout, and a few hairs are present on both jaws and the rostrum.

In dorsal profile the body is slim and streamlined. There are 32–60 ventral grooves, all ending well before the navel. The flippers are relatively small, about one-eleventh the body length, and pointed on the tips; the flukes are small relative to overall size.

The rather erect dorsal fin, which is usually from 25 to more than 60 cm tall and strongly falcate in adults, is located just more than one-third of the body length forward from the fluke notch, slightly farther forward than that of the fin whale and much farther forward than that of the blue whale. It often has a small "knee" about halfway up the forward margin.

Sei whales are dark gray on the back and sides and on the posterior portion of the ventral surface. The body often has a galvanized appearance because of scars that may be caused by lamprey bites inflicted during migrations into warmer waters. These scars may be dark gray or almost white. On the belly there is a region of grayish white that is confined to the area of the median ventral grooves. Neither the flippers nor the flukes are white underneath. The right lower lip and the mouth cavity, unlike those of the fin whale, are usually uniformly gray.

The baleen plates, of which there are about 300 to 410 per side, are uniformly ash-black with fine white fringes. The texture of the fringes is almost silky, having been likened to sheep's fleece. A few sei whales have several half-white plates near the front of the mouth, and so might be mistaken for fin whales.

NATURAL HISTORY: Sei whales usually travel in groups of two to five individuals, though they may concentrate in larger numbers on the feeding grounds. Every second or third year, usually in winter and after a gestation period of approximately twelve months, a female gives birth to a single calf, which she nurses for five to nine months. At weaning, calves are 8 to 9 m long. Average age at sexual maturity in some areas of the North Pacific reportedly declined from ten years in 1930 to six for females and seven and a half for males by the early 1960s, but in less exploited areas remained at ten years for both sexes. In the Antarctic, average age at sexual maturity similarly declined from ten or eleven years in 1930 to eight by 1963. Responses of this kind—in which various aspects of reproductive strategy change as a function of population density—have been described for other baleen whales. However, the notion that increased productivity results from controlled reduction in population size (i.e., the more you kill, the more you can harvest) is currently a subject of heated debate.

Like other baleen whales, sei whales undertake annual migrations from lower-latitude wintering grounds to higher-latitude feeding grounds, though they may depart the winter grounds later than, and generally do not extend into as high latitudes as, some other balaenopterids. Movements of sei whales are thought to be unpredictable—off Korea, Norway, and Japan, and in the western North Atlantic—and researchers have reported "invasions" in irregularly spaced years. This irregularity, if it is real, would probably result from changes in environmental conditions, which render an area more or less productive from year to year.

The blow of sei whales is an inverted cone, much like the fin whale's but lower and less dense. Sei whales are primarily skimmer feeders, and usually do not dive very deeply. For that reason, they usually surface at a shallower angle than fin and Bryde's whales. The head rarely emerges at a steep angle, except when the whales are being chased. Instead, the blowholes and a major portion of the back, including the dorsal fin, become visible almost simultaneously, and remain visible for relatively long periods. When they begin another dive, sei whales do not arch the tail stock or flukes high. Instead, they normally submerge by slipping quietly below the surface, often remaining in view only a few meters down and leaving a series of tracks or swirls on the surface as they move their flukes. When feeding in this manner, sei whales may exhibit a highly regular blowing and diving pattern for long periods. They were regarded by early whalers as the fastest swimmers of the rorquals.

In northern portions of their range, sei whales feed primarily on copepods. Throughout the remainder of the range, however, their food is more varied, and regularly includes krill (euphausiids), squid, and a variety of small schooling fish.

DISTRIBUTION AND CURRENT STATUS: Sei whales are distributed

worldwide, and though their distribution and movements during much of the year are poorly known, the species appears to favor temperate and oceanic waters. Their presence in some parts of their range is unpredictable and sporadic. In general, they do not venture as far poleward as fin whales, but may have a greater tendency to enter tropical waters. One problem in mapping their distribution in the tropics is the difficulty of distinguishing them at sea from Bryde's whales.

In the eastern North Pacific, sei whales are found in summer from central California north through the entire Gulf of Alaska; at least some of those off California migrate to the waters off British Columbia. They winter from at least 35°30′ N (Piedras Blancas, California) south to 18°30′ N, near the Revillagigedo Islands off Mexico. In the western North Pacific they are common off Japan and Korea, north to the southwestern Bering Sea. They can be found offshore in a broad arc between about 40° N and 55° N, continuing uninterrupted across the Pacific. The North Pacific population, which comprises three putative stocks, probably numbers more than ten thousand whales, but it is considered depleted and thus is completely protected.

In the western North Atlantic, two stocks of sei whales are recognized, one centered on the Nova Scotian shelf, numbering perhaps one or two thousand, the other in the Labrador Sea, comprising about one thousand whales. Individuals are known to wander to the West Greenland coast, and sightings as far south as Florida are probably of winter migrants from these northern stocks. Most reports of sei whales in the Gulf of Mexico and Caribbean Sea are questionable because of possible confusion with Bryde's whales.

Sei whales in the eastern North Atlantic are thought to winter off Spain, Portugal, and northwest Africa, and to migrate north to northern Norway, Bear Island, and Novaya Zemlya, occasionally reaching Spitsbergen. Though they were once abundant in these areas, there are no recent estimates of population size, and the stocks are believed to be depressed. Sei whales are still hunted in summer in Denmark Strait by Iceland, but the winter whereabouts of these whales is not known.

In the southern hemisphere, movements of sei whales reportedly are similar to those of fin whales, but cover a smaller range. Tag returns have linked sei whales from separate antarctic zones with grounds off Natal, South Africa, western and southeastern Australia, and Brazil. They are found primarily south of latitude 30° S in winter and south of 40° S in summer, though they are common only around South Georgia in late summer and autumn. Southern hemisphere stocks of sei whales, once numbering perhaps 150,000 to 225,000, have been heavily exploited, and have suffered major declines in all areas. Because they are versatile feeders, sei whales are believed to have increased in numbers as the populations of blue, fin, humpback, and right whales were reduced.

Researchers Bernd and Melanie Würsig measure a sei whale that washed ashore near where they were studying dusky dolphins. (Golfo San Jose, Argentina, December 1973: courtesy of Bernd Würsig.)

CAN BE CONFUSED WITH: The sei whale's smaller size and decidedly taller, more falcate dorsal fin, located farther forward than those of the other large balaenopterids, should readily prevent confusion with the blue whale, and may help distinguish it from the fin whale and the Bryde's whale when viewed closely. At a distance, however, sei whales are hard to tell from fin and Bryde's whales; so the observer must patiently accumulate multiple evidence before arriving at a decision. The primary clues for distinguishing sei whales from fin whales are discussed in the section on fin whales.

Sei whales may be distinguished from Bryde's whales only by close examination. The dorsal fin of Bryde's whales is usually smaller, less than 45 cm high, is sharply pointed, and is often worn on the rear margin. It usually is located in approximately the same region of the back as that of the fin whale, perhaps slightly farther aft than the fin of the sei whale. Bryde's whales are primarily fish eaters, and they dive more like fin whales than sei whales. When close enough, one can see that the sei whale has only a single head ridge, but Bryde's whales have two additional ridges, one on each side of the main ridge.

Bryde's Whale

Balaenoptera edeni
Anderson, 1878
DERIVATION: *edeni* refers to the
Honorable Ashley Eden, Chief
Commissioner of British
Burma, who secured the type
specimen for Anderson.

ZONES 1 TO 5

DISTINCTIVE FEATURES: Head with a series of three prominent ridges from the area of the blowholes to the snout; body dark gray overall; baleen slate gray, with coarse, dark bristles; primarily found in subtropical and tropical regions.

DESCRIPTION: Two forms exist in some areas, one coastal and one offshore. Maximum body length is about 14 m. Females are larger than males, and attain sexual maturity at a slightly longer average length, 12.5 m versus 12.2 m. Offshore forms are slightly larger than coastal forms.

Bryde's whales closely resemble sei whales externally, and the two species were long confused. Until recently, data on catches and biology of the two species have been summarily treated as if all were sei whales. Bryde's whales are smaller, however (21 versus 14 m, maximum length). Their heads may be similar in profile and general appearance at a distance, but the two species can be distinguished at close range by head shape. Most Bryde's whales have three prominent ridges on the head anterior to the blowholes, one medial, and one on each side from the area adjacent to the blowholes toward the tip of the snout. (As is true for many other features of balaenopterids, there are numerous intergrades of this feature, from prominent to subdued or not present at all.) If the ridges are present and can be examined at close range, they are sufficient grounds by themselves to allow Bryde's whales to be positively identified.

Bryde's whales reportedly have forty to fifty ventral grooves that extend at least to the navel. The dorsal fin can be nearly 46 cm tall; it is extremely falcate and pointed on the tip. It is located in about the same position as that of the fin whale, usually about one-third of the body length forward from the fluke notch. Bryde's whales are generally dark gray dorsally, though some individuals have regions of light gray on each side forward from the dorsal fin (somewhat like minke whales); on the back between the head and the dorsal fin (observed in Venezuelan juveniles); or in a band along the side. The belly and chin are reportedly white.

The species has 250 to 370 slate-gray baleen plates, up to 42 cm long and 24 cm wide, usually with light gray bristles occurring in a density of 15 to 35 per square centimeter. Plates have also been reported with fine white bristles. The baleen plates of the offshore form are longer and broader than those of the coastal form, and closely resemble those of sei whales.

NATURAL HISTORY: Bryde's whales usually occur singly or in small groups. Concentrations have been noted in several areas off Japan, Peru, Brazil, South Africa, and Venezuela, and in the Gulf of California.

Females usually give birth every second year. Gestation lasts about one year. Mating and calving may occur year-round, but in the northern hemisphere, at least, calving peaks in autumn. The offshore population off South Africa breeds in winter, the inshore population breeds year-round.

Some tropical populations may be sedentary, but those in temperate regions are thought to be migratory. Bryde's whales are more abundant off the Venezuelan coast from late spring and summer through December than during the rest of the year, suggesting seasonal migrations there.

Bryde's whales, like minke whales, reportedly often approach close to vessels as if curious about them. During this time they may be examined carefully and their identifying characteristics noted.

Though euphausiids may be an important component of this species' diet in some areas, both coastal and offshore Bryde's whales appear to prefer schooling fish, including pilchards, anchovies, herring, and mackerel. They have been observed off Venezuela feeding on bonito. Overall they are thought to feed higher in the food chain than sei whales, a likely explanation for observed differences in the diving behavior of the two species. Bryde's whales are not skimmer feeders like sei whales; they are deeper divers, and are more animated in the actions that are visible to a surface observer.

When they surface to breathe, they often rise steeply to the surface, exposing much of the head, roll the body sharply, and hump the tail stock before beginning another dive. When feeding, Bryde's whales often roll onto their sides or churn the water at the surface by pinwheeling or halfheartedly breaching. In this mode they closely resemble fin whales. They apparently do not raise the flukes when beginning a dive. Unlike sei whales, Bryde's whales frequently accelerate and change direction suddenly; so their movements during feeding look more like those of feeding dolphins than like those of other large whales.

DISTRIBUTION AND CURRENT STATUS: Bryde's whales are found in tropical and warm temperate waters around the world, often near shore in areas of high productivity. Little is known about the population(s) in the Atlantic. On the western side, they are found from off the southeastern United States and the southern West Indies (with apparently resident populations in the Gulf of Mexico and the southeastern Caribbean) to Cabo Frio, Brazil. Abundance of these populations is unknown, though the Venezuelan stock may contain more than a hundred individuals. On the eastern side they range from the vicinity of the Strait of Gibraltar south past the Cape of Good Hope.

In the Indian Ocean, Bryde's whales range from the Cape of Good Hope north to the Persian Gulf, east to the Gulf of Martaban, Burma, and thence south to Shark Bay, Western Australia.

In the western Pacific they are distributed from northern Hokkaido, Japan, south to Victoria, Australia, and North Island, New Zealand; in the eastern Pacific they have been reported from southern California south to Iquique, Chile. They are present in the equatorial Pacific, and here, as in other areas, the degree of mixing between northern- and southern-hemisphere stocks is unknown.

The smaller coastal form is found within twenty miles of the coasts of South Africa and western Kyushu, Japan, off Brazil, and probably off Baja California. Bryde's whales from all other areas, including offshore waters of Brazil, Japan, and South Africa, are of the offshore form: larger, more heavily scarred, with longer and broader baleen plates.

Known harvests of Bryde's whales were rare prior to 1920; but since then, partly because of a greater effort to exploit Bryde's whales knowingly, and partly because of increased awareness of their differences from sei whales, recorded harvests have increased, particularly in the western North Pacific and off Peru. Even tentative estimates of population size are available only for the western North Pacific, where a population once estimated at 21,000 or more apparently has been reduced to about 14,000.

CAN BE CONFUSED WITH: At sea Bryde's whales can be confused

The serpentine carcass of a stranded Bryde's whale, showing the three head ridges characteristic of the species. The mouth holds a pool of water that reflects the left row of baleen. (Fort George Island, Florida, March 14, 1978: Marineland of Florida.)

with sei whales, fin whales, and perhaps minke whales. When examined at close range, most Bryde's whales can be distinguished from other balaenopterids by their three prominent head ridges. All others have a single head ridge. Some fin whales have lateral ridges, but they are usually much less prominent than the median ridge.

Bryde's whales might be distinguishable from sei whales by differences in diving behavior. The shallow-feeding sei whales usually surface and blow at regular intervals for long periods of time. Bryde's whales are usually deeper divers, less likely to surface and blow at evenly spaced intervals. If they are seen only briefly or at a distance, however, the two species may be impossible to differentiate. When a specimen is available, Bryde's whales can most often be distinguished from sei whales by differences in the length of the ventral grooves, and from fin, sei, and minke whales by differences in number, dimensions, color, and texture of the baleen plates.

During winter months, fin whales venturing into tropical waters might be mistaken for Bryde's whales. However, the head of the fin whale is often more acutely pointed. Furthermore, its right lower lip and the right front baleen are white. The baleen and the right lower lip of Bryde's whales are dark gray. If the animals can be approached closely from the right side, fin whales can be identified or ruled out by these color differences.

Like Bryde's whales, minke whales often approach vessels, but minke whales have an acutely pointed snout, a single head ridge, and, in most populations, a white band on each flipper. Furthermore, minke whales rarely reach 10 m in length; Bryde's whales reach 14 m.

Minke Whale

Balaenoptera acutorostrata
Lacépède, 1804
DERIVATION: from the Latin
acutus for "sharp, pointed," and
rostrum for "beak, snout."

ALL ZONES

DISTINCTIVE FEATURES: Head medially ridged and sharply V-shaped when viewed from above, all dark; tall, falcate dorsal fin, positioned in aft third of back, usually appears simultaneously with low, indistinct blow; dorsally black or dark gray, ventrally white; area of gray shading extending up each side in front of and below dorsal fin; flippers have transverse white band in northern hemisphere; flipper band absent or irregular in position and shade in southern hemisphere; ventral grooves end anterior to navel; baleen yellowish white anteriorly, becoming gray to brown-black posteriorly.

DESCRIPTION: This is the smallest member of the genus *Balaenoptera*. Northern- and southern-hemisphere minke whales differ in several ways, and are currently considered to include two or three subspecies. Maximum length is about 10.7 m for the female and 9.8 m for the male, with a weight of ten tons in the southern hemisphere; a length of 9.2 m is maximum in the northern hemisphere. Sexual maturity is reached at an estimated age of seven to eight years, at which time the animals are 6.7 to 7 m (males) and 7.3 to 7.9

m (females) in the northern hemisphere, and 7.2 to 7.7 m (males) and 7.9 to 8.1 m (females) in the southern hemisphere. Newborn are 2.4 to 2.8 m.

One of the most distinctive features of the species is the extremely narrow, pointed, and triangular rostrum with a single prominent head ridge, similar to but much sharper than that of the fin whale.

The body is slender and strikingly streamlined. The flippers are slim and pointed on the tips. The flukes are broad, to about one-fourth the body length, and smoothly concave along the notched rear margin. There are fifty to seventy ventral grooves ending anterior to the navel, often just behind the flippers.

Minke whales have a tall, falcate dorsal fin, located nearly two-thirds of the way back from the tip of the snout. It often becomes visible simultaneously with the low, usually inconspicuous blow.

Minke whales are black to dark gray on the back, and white on the belly and on the underside of the flippers. Portions of the underside of the flukes may be bluish steel gray. In the northern hemisphere they have a diagonal band of white on each flipper, the extent and orientation of which varies individually. This band is less frequently present in the southern hemisphere, and varies geographically in its shape and intensity.

Minke whales sometimes have a chevron on the back behind the head, and they often have two regions of light gray on each side, one just above and behind the flippers, another just in front of and below the dorsal fin.

There are 230 to 360 (generally about 300) baleen plates. Atlantic minke whales apparently have more plates than Pacific minke whales. Plates reach maximum dimensions of 12 cm at the base and just over 20 cm in length, and in the northern hemisphere are mostly yellowish white, with fine, white bristles. Some of the white plates have black streaking, and the posteriormost plates can be all dark. In southern-hemisphere minke whales, the baleen is often dark grey; the anteriormost plates can be white; and the gray plates frequently have white inner edges.

NATURAL HISTORY: Minke whales are frequently found as single animals, pairs, or trios, though they may congregate in areas of food concentration in polar seas during spring and summer. Schools or groups of several hundred have been reported from the Antarctic. In all areas minke whales appear to segregate by age/sex classes more than any other baleen whale, a factor that has hindered attempts to use samples from harvested animals to obtain unbiased estimates of population size and composition. In Puget Sound, Washington, sixteen individually identifiable minke whales maintained exclusive adjoining ranges during the summer they were observed. Some minke whales undertake lengthy migrations, totaling 9,000 km or more, but others may move little; there are many

Minke whales may appear in almost any marine habitat: from near shore to blue pelagic, from tropics to polar ice. (Spieden Channel, San Juan Islands, Washington, October 1981: Betsy Rupp Fulwiler.)

regions where minkes are found year-round. They are more likely to be seen up close than their larger cousins (the blue, fin, and sei whales) because they often closely approach boats, particularly stationary ones, as if curious about them.

Minke whales approach close to shore and often enter bays, inlets, and estuaries. Like fin whales, they often arch the tail stock high above the surface when beginning a long dive. However, they do not raise the flukes above the surface. Minke whales sometimes breach, leaping completely clear of the water and entering either smoothly and head first, or with a substantial splash like humpback whales.

Minke whales feed primarily on krill (euphausiids) in the southern hemisphere and on small shoaling fish (herring, cod, pollack, and capelin) and krill in the northern hemisphere. They also eat copepods occasionally in all areas. Minke whales tend to feed on whichever preferred food source is most abundant in a given area; so their apparent preference for krill in the southern hemisphere, which is based on samples from the Antarctic, may result from the fact that the Antarctic lacks the large concentrations of the schooling fish abundant in the northern hemisphere.

These small whales are themselves preyed upon with surprising frequency by killer whales, particularly in certain parts of the southern hemisphere. Several have been held for periods of one to three months in captivity in Japan; the minke is the only member of its genus small enough to be considered seriously for captive display and research.

DISTRIBUTION AND CURRENT STATUS: Minke whales are widely

distributed in tropical, temperate, and polar waters of both hemispheres. Like other balaenopterids, they make their most equatorial penetration in winter, their closest approach to the ice in summer, though overall the species appears to be widely distributed in all seasons, and to migrate in a manner hard to predict from year to year. At least three geographically isolated populations are recognized: North Pacific, North Atlantic, and southern hemisphere. These may well comprise local stocks that are genetically isolated, especially in the southern hemisphere.

In the North Pacific the majority winters from central California (in the east) and 40° N (in the west) to near the equator, with major concentrations between 20° N and 25° N. In summer they shift northward, ranging from northern Baja California (in the east) and 35° N (in the west) through the Bering Sea, with stragglers ranging as far as Barrow, Alaska, in the Chukchi Sea. During this season they may be found throughout much of the North Pacific Basin, though few sightings have been recorded within approximately 600 km of the Hawaiian Islands.

Historically, North Pacific minke whales have been killed by shore-based fisheries in Korea and Japan. The Korean fishery is now the largest coastal minke-whale fishery in the world. The modern incarnation of the centuries-old Japanese fishery operates with small shore-based vessels, taking a few hundred whales each year. Minke whales are not taken by pelagic whaling vessels in the North Pacific. Individuals are taken sporadically for food by Eskimos on St. Lawrence Island, Alaska.

This curious minke whale approached a research vessel, circled it repeatedly, then vanished. (About 13 km off Dana Point, California, April 1982: Brent S. Stewart.)

In the North Atlantic, minke whales are distributed from the Lesser Antilles and eastern Gulf of Mexico in the west, and the central and western Mediterranean Sea in the east, northward to the edges of pack ice. In winter they appear to be most abundant in temperate waters across the entire North Atlantic, infrequently entering tropical waters. In summer, concentrations shift northward to Spitsbergen and the Barents Sea, the coast of Norway, and waters off Iceland, Greenland, and Newfoundland. Norwegian catcher boats, working coastal and pelagic regions alike, killed more than 52,000 minkes between the early 1920s and 1973, and an average of about 2,000 per year from 1969 through 1975. Scores of minkes are killed off Iceland in some years, and in western and southern Greenland two fisheries—one using small commercial catcher boats, the other consisting of natives in small outboard-powered boats and using rifles and hand harpoons—reportedly take more than 200 a year.

Southern-hemisphere distribution of the minke has been described as pelagic and circumpolar, with recorded sightings from the pack ice to the tropical South Atlantic, Indian, and South Pacific Oceans. In the South Atlantic their migrations have been described as annual oscillations about 42° S, in which larger and older animals (large males and pregnant females) range farther south to the ice edge, but smaller animals (nonpregnant females, calves, and immature whales) remain in waters between about 50° S and 20° S.

Minkes are common off Durban, South Africa, where a modern shore-based fishery took roughly 100-200 per year between 1968 and 1976. They reportedly are rare in the waters off

Researchers on shore-fast ice record underwater sounds from a minke whale against the backdrop of 3,794 meter Mount Erebus, on Ross Island. (McMurdo Sound, Antarctica, January 1981: Stephen Leatherwood.)

western South Africa, but do range as far north along the West African coast as 5° S, off Angola. On the opposite side of the South Atlantic, they also range to 5° S, and enough inhabit coastal shelf and slope waters off Brazil to support a shore fishery based at Costinha (7° S). That fishery has, since the mid-1960s, taken about 600–1,000 minkes per year.

Elsewhere in the southern hemisphere, minke whales are the only baleen species still taken by pelagic factory-ship whaling. Since the 1971–1972 season, they have replaced in importance the larger, more depleted species, with total catches during the 1970s ranging as high as 8,100. Most whaling for minkes is done for meat for human consumption, with oil and meal generally regarded as by-products.

CAN BE CONFUSED WITH: When seen at relatively close range, minke whales can be distinguished readily from the other rorquals that have relatively tall, falcate dorsal fins (fin, sei, and Bryde's whales) by the combination of their much smaller size, acutely pointed head with dark face and lips, and, when present, the distinctive white band on each flipper.

At a distance, however, positive identification may be difficult, and care should be taken. Minke whales have a low, inconspicuous blow. Like sei whales, they frequently expose the dorsal fin simultaneously with the blow, but minke whales usually hump the tail stock much higher when beginning a long dive, more like fin whales.

From a distance minkes might also be mistaken for bottlenose whales (or any of several other beaked whales with a similar dorsal fin) and pygmy right whales (southern hemisphere only). The beaked whales have slimmer heads (bulbous in the bottlenose) and some form of beak, whereas the minke's head is sharply triangular. As observations of free-ranging beaked whales have increased, it has become clear that some previous reports of minke whales were actually of beaked whales. Although there have been only a few recorded encounters with live pygmy right whales, they appear to have a much more arched rostrum and a correspondingly contoured mouthline. It is not clear whether pygmy right whales have also been misidentified as minke whales by careless observers.

Humpback Whale

Megaptera novaeangliae
(Borowski, 1781)
DERIVATION: from the Greek
megas for "large," and *pteron* for
"wing" or "fin"; from the Latin
novus for "new," and the
Middle English *angliae* for
"England." ALL ZONES

MOST DISTINCTIVE FEATURES: Flippers very long (to nearly one-third of body length), often white or partly white, with knobs on leading edge; head in front of blowholes flat and covered with knobs; dorsal fin can have various shapes, is often stepped or humped; flukes and flippers scalloped on trailing edge; flukes often raised on dive.

DESCRIPTION: Female humpback whales can be 16 m long; males, 15 m. Females attain sexual maturity at 11.4 to 12.4 m, males at 11 to 12 m. Newborn are 4.5 to 5 m long.

Viewed from above, the head is broad and rounded, somewhat like that of the blue whale, though shorter. The median head ridge characteristic of balaenopterids is indistinct, and is marked by a string of fleshy knobs or protuberances, many more of which are distributed elsewhere on the head and lower jaw. In profile, the head is surprisingly slim, almost alligator- or gar-like in appearance. There is a rounded protuberance near the tip of the lower jaw.

The body is more robust than those of other balaenopterids. There are about 14 to 35 broad ventral grooves, to 38 cm wide, extending at least to the navel. The flippers are very long (nearly one-third as long as the body), with knobs or bumps on at least the leading edge. They are nearly all white in some populations.

The dorsal fin is located a little less than one-third of the body length from the fluke notch, in approximately the same position as that of the fin and Bryde's whale. It can range in size and shape from a small, triangular nubbin to a more substantial, sharply falcate fin. The dorsal fin frequently includes a step or hump, which is accentuated when the animal arches its back to begin a dive and from which the species derives its common name. The flukes are broad and but-

Water spews from the mouth of a lunge-feeding humpback whale. (Glacier Bay, Alaska, August 1975: Ronn Storro-Patterson.)

terfly shaped, and marked on the rear margin by serrations and frequent irregularities.

Humpback whales are basically black or gray, with a white region of varying extent on the throat and belly; the flippers are white underneath and sometimes above as well; the undersides of the flukes can be completely or partially white. The 270–400 baleen plates on each side, the largest of which is up to 70 cm long and 30 cm wide, are generally all black, with black or olive-black bristles.

NATURAL HISTORY: Humpback whales are often found alone or in groups of two to three, but throughout their breeding and feeding ranges may congregate in groups of up to twelve or fifteen. Small feeding groups may assemble in areas of food abundance.

Although females seem occasionally to give birth in successive years, they normally reproduce at intervals of two or more years. Following a twelve-month gestation period, adult females give birth to a single calf, which they nurse for ten to eleven months. The calf is about 8 to 9 m long when weaning is completed.

Humpbacks make extensive seasonal migrations between areas of concentration on high-latitude summering grounds and low-latitude wintering grounds, the latter along continental coasts or around oceanic islands. Their distribution in general is over shallow banks and in shelf waters.

The humpback's blow is 2.5 to 3 m tall, and bushy. It is generally wide relative to its height, though when surfacing from a long dive, the humpback, like other balaenopterids, may emit a more ambitious blast, taller and more slender than usual. Feeding humpback whales habitually blow four to eight times at intervals of fifteen to thirty seconds after a long dive. In the tropics they often blow only two to four times before beginning a long dive. As they dive, humpback whales frequently throw their flukes high into the air, exposing the white or partially white undersurface and the rippled rear margin. Dives of twenty minutes have been reported, though dives lasting more than fifteen minutes seem to be rare.

Humpback whales are not particularly fast swimmers, generally traveling between 6 and 12 km per hour. Nevertheless, they often muster enough energy and speed to leap clear of the water in a dramatic and unexpected marriage of power and grace. There is no large whale species more animated or acrobatic than the humpback. They often raise a flipper and slap it against the water, or "lobtail," raising the tail high into the air and bringing it crashing back to the water in a loud report. Often a humpback whale will be found lying on its side with a long flipper in the air or on its back with both flippers in the air.

A humpback whale breaching in Glacier Bay, Alaska. (Al Giddings/Ocean Films Ltd.)

Humpback whales feed on krill and schooling fish. One interesting feeding behavior has been described in which the animals circle a school of fish or krill from below, emitting a bubble curtain as they ascend slowly to the surface. Once fish or plankton are confined within this "bubble net," the whales charge through it, their mouths open, engulfing the prey. Some evidence shows that during such feeding there may be mandibular kinesis, an actual unhinging of the mandible, as is known for many reptiles.

Barnacles and "whale lice" (cyamid crustaceans) find the slow-moving humpback's skin an acceptable home and frequently infest it. Humpbacks are vulnerable to attacks by killer whales, and they sometimes become fouled in fixed fishing gear, particularly in inshore areas of New England and eastern Canada, where the species is locally and seasonally common.

DISTRIBUTION AND CURRENT STATUS: Humpback whales are widely distributed in all oceans, ranging from their tropical wintering grounds around islands and continental coasts to the edges of polar ice zones, though they generally do not go to the pack-ice edge.

In the North Pacific, presumably discrete stocks winter around the Mariana, Bonin, and Ryukyu Islands and Taiwan; around the main Hawaiian Islands; and along the mainland American coast and nearshore islands, from central Baja California to Cabo San Lucas, from southern Sonora to Jalisco, Mexico; and summer from the coasts of Honshu and southern California north into the Chukchi Sea, chiefly near shore. The extent to which these stocks mix on the summer grounds is poorly known.

In the western North Atlantic, humpback whales visit Bermuda on migration, and winter in the West Indies south to Trinidad and northeastern Venezuela. In the eastern North Atlantic, they winter around the Cape Verde Islands and off northwest Africa. These stocks summer from New England north to southeastern Baffin Island, along the west coast of Greenland, north to Disko Bay, and around Iceland; and from northern Norway northward.

In the southern oceans five (perhaps six) largely isolated stocks summer in Antarctic waters. They migrate annually along predictable routes to winter in waters off Brazil; off West Africa; off Madagascar; off western Australia; in the Coral Sea; near New Zealand to the Tonga Islands; and off Chile, Peru, and Ecuador as far north as the Galapagos Islands.

Because of their tendency to concentrate on both summer and winter grounds, often near coasts in easily accessible areas, humpbacks were easy prey for shore-based whalers and were so severely depleted everywhere that recovery has been extremely slow. Some southern-hemisphere stocks do not seem to have increased appreciably since they were afforded complete protection in 1964. A few individuals are still taken each year in "subsistence" fisheries in western Greenland and the Lesser Antilles.

Of perhaps 15,000 humpbacks that existed in the North Pacific at the onset of mechanized commercial whaling, less than 1,000 may survive, most of them on the eastern side (including Hawaii). The western North Atlantic stock is relatively healthy, containing upward of 2,000 whales, but the eastern North Atlantic stock is in relatively poor shape. In the southern oceans, there were probably close to 100,000 humpbacks in the nineteenth century; today there are believed to be only about 2,500.

CAN BE CONFUSED WITH: From a distance, humpback whales can be confused with any of the other large balaenopterids: blue, fin, sei, or Bryde's whales. Although the humpback's dorsal fin can differ greatly between individuals, it most closely resembles that of the blue whale. However, it is located farther forward on the back, and is generally more prominent. Some humpbacks have a dorsal fin indistinguishable from that of a fin or sei whale. Unlike other rorquals, humpbacks have a habit of raising the flukes, which are shaped uniquely on the rear edge, high into the air when starting a long dive. (In shallow water they may not raise the flukes at all.) The only other rorqual to do so, the blue whale, raises the flukes slightly or not at all. At close range the knobs on the humpback's head and its long, often white or partially white flippers are diagnostic.

Under some conditions, humpback whales may be confused at a distance with sperm whales. When arching the back to begin a dive, both may show a distinct hump. Both species frequently raise their flukes nearly vertically when beginning a long dive, but the flukes differ in several ways. The humpback's flukes show varying amounts of white on the undersides, and are distinctly concave and irregularly rippled on the rear margin. Those of sperm whales are all dark, and more nearly straight and even along the rear margin.

The right whale often lifts its flukes vertically, too, and this can lead to confusion with the humpback. Right-whale flukes are even along the rear margin and all black. The bushy blow of the humpback usually differs from the sperm whale's, which emanates from the front of the head and is angled forward and to the left, and from the right whale's, which is distinctly V-shaped.

3. The Gray Whale

Family: Eschrichtiidae

Gray Whale

Eschrichtius robustus
(Lilljeborg, 1861)
DERIVATION: *Eschricht* refers to
a nineteenth-century Danish
zoology professor; from the
Latin *robustus* for "oaken" or
"strong."

ZONES 1 AND 7

DISTINCTIVE FEATURES: Body mottled gray; head narrow, V-shaped when viewed from above, and bearing barnacles and "whale lice"; dorsal fin absent; bumps or ridges on top of tail stock; blow low and heart-shaped; distribution, North Pacific; primarily coastal.

DESCRIPTION: Maximum body length is approximately 14.1 m. Females are generally larger than males of the same age. Larger whales may weigh 2 to 2.5 tons per meter, reaching maximum total weights of more than 35 tons. Sexual maturity is reached at about 11.1 m in males and 11.7 m in females, between five and eleven years of age. Newborn calves average approximately 4.9 m in length.

The gray whale's head is narrow and triangular dorsally and laterally, with an arched upper contour in lateral profile. It is often encrusted with barnacles. Some clusters of barnacles, and most indentations or creases of the skin, are inhabited by yellowish white to yellowish orange cyamid crustaceans usually referred to as "whale lice." The contour of the mouth is slightly arched. The top of the tapering upper jaw has many pits or depressions, most containing small hairs believed to link to sensory fibers. The mouth contains 130–180 relatively small (5 to 25 cm) yellowish white baleen plates per side. The plates have coarse bristles, which are also yellowish white.

Instead of the longitudinal ventral grooves found in balaenopterids, there are two to five deep creases on the throats of gray whales. There is no dorsal fin, but instead a low hump followed by a series of six to twelve knobs or crenulations of varying emphasis along the dorsal ridge of the tail stock (often referred to as "knuckles"). As the name suggests, the body is robust, particularly in the region of the flippers.

Gray whales are mottled gray, though the darkness of the base and the extent of the blotching may vary. When barely submerged, they may appear to a surface observer to be marbled white or pale blue. Calves are generally more uniformly dark, and older animals exhibit greater color. The body, especially the dorsal surface, is varyingly covered with patches of barnacles and associated "whale lice." The orange or yellow "whale lice" are also found on grooves and folds, and on the open surface of the skin. The flippers, with rounded margins and pointed tips, and the broad flukes, over 3 m wide in large adults, are mottled.

A gray whale lunges to the surface, creating a substantial pressure wave. (San Ignacio Lagoon, February 1979: Steven L. Swartz.)

NATURAL HISTORY: The gray whale's natural history is dominated by, and best told within the framework of, its annual migration, one of the longest known for any mammal. The vast majority of the population spends from about May through November feeding on benthic amphipods, which abound in parts of the Bering, Chukchi, and western Beaufort Seas. They spend the remainder of the year on migration and in temperate and tropical breeding lagoons off Baja California and mainland Mexico. Migrating whales are seen in groups of as many as sixteen, though singles, pairs, and trios are most common. There is good evidence that the composition of migrating pods that do not contain any females with calves is in constant flux. In and near breeding lagoons, female-calf pairs, and groups of consorting adults and juveniles, are most common. On northern feeding grounds, they are often encountered singly, though up to several hundred may be present in patchily distributed food-rich areas a few dozen kilometers square.

The birth season usually lasts from January through March. Females bear a single calf, probably at intervals of two or more years. They defend the calf aggressively, a characteristic that earned them the name "devilfish" from Yankee whalers, who often harpooned young to keep mothers within range. Weaning appears to occur within nine months, certainly by the fall migration.

Migrating gray whales exhibit predictable respiratory patterns, generally blowing three to five times at intervals of fifteen to thirty seconds before raising the flukes and submerging for three to five

A gray whale seen from the crow's nest of a Soviet research vessel. (Off Cape Lisburne, Alaska, 67° 37' N, 166° 10.8' W, August 7, 1982: David Rugh, courtesy of U.S. National Marine Fisheries Service.)

minutes. This highly regular cycle is repeated again and again on both the southbound migration (on which the whales swim at about 6 to 8 km per hour) and the northbound migration (on which adults swim at about 6 to 8 km per hour, females with calves at 3 to 5 km per hour). The blow is generally low (less than 3 to 4 m) and puffy, described by some as heart-shaped on windless days. Behavior on the feeding and calving grounds is more erratic than during migration, with whales surfacing and diving at variable intervals of up to fifteen minutes. Throughout their range, but most apparently in lagoon concentrations, gray whales breach and raise their heads vertically out of the water (spy-hopping or pitch-poling).

In the north, feeding whales frequently surface with clouds of mud and detritus streaming from the sides of the mouth, a meal-call for hungry birds that attend them. Similar feeding behavior has been observed in other parts of their range, and has led some investigators to believe that gray whales feed opportunistically year-round, although direct evidence is lacking. In most of their range, gray whales feed almost exclusively on benthic gammarid amphipods, organisms that appear to thrive in disturbed communities of the kind left behind by bands of gray whales after bottom-grubbing for food. Gray whales present year-round off Vancouver Island are known to feed on dense epibenthic mysid concentrations.

Killer whales may be a significant cause of gray-whale mortality, especially among young animals on migration. There are numerous reports of killer whales feeding on the tongues of gray whales, leaving the carcasses as carrion. Living gray whales are often seen with scars caused by teeth of killer whales. Several species of large sharks, abundant in and around the calving lagoons, scavenge whale carcasses and may account for some calf mortality.

DISTRIBUTION AND CURRENT STATUS: Three or more stocks of gray whales are recognized: North Atlantic stock (or stocks), apparently exterminated in recent historical times (perhaps as late as the seventeenth or eighteenth century); a Korean or western Pacific stock, hunted at least until 1966, and now extinct or very close to extinction; and a California or eastern Pacific stock.

The California stock is the one whose annual migration, described in the preceding section, supports a very significant whale-watching "industry" along the west coast of North America during November through March. The principal gray-whale wintering grounds are Scammon's Lagoon (Laguna Ojo de Liebre), Black Warrior Lagoon (Laguna Guerrero Negro), San Ignacio Lagoon (Laguna San Ignacio), and the Magdalena Bay complex (from Boca de las Animas to Bahia Almejas), all on the outer coast of Baja California. The migration, especially in the spring, is primarily coastal (except in the Southern California Bight and Gulf of Alaska), most

Researcher Gayle Dana touches the head of a "friendly" lagoon gray whale. (San Ignacio Lagoon, March 1979: Steven L. Swartz.)

whales remaining inside the 100-fathom curve. They are frequently seen in or near the surf, and have been reported swimming into lagoons, bays, and river mouths. The northward migration occurs as two distinct phases, separated by about eight weeks. The second consists primarily of females with their calves, lagging behind the earlier migrating portion of the population. Recently, small bands have been observed during all seasons of the year near the Farallon Islands, off Point St. George, off the Klamath River mouth, and off Big Lagoon, all in California; off Oregon; and off Vancouver Island and the Queen Charlotte Islands, British Columbia.

In spring, gray whales funnel into the Bering Sea through Unimak Pass, press close along the shore until well inside Bristol Bay, and then turn northward toward Nunivak Island (small numbers also travel to the Pribilof Islands). From Nunivak Island and St. Matthew Island, they fan out northward across the open Bering Sea, avoiding deeper waters of the southwestern Bering Sea and coastal shoal areas north of Nunivak Island, including Norton and Kotzebue Sounds. As the ice recedes, they press northward, the most adventuresome wanderers reaching at least Barter Island near the United States-Canadian border and Wrangel Island off the Soviet coast. As ice advances at summer's end, they reverse the process, again streaming through Unimak Pass.

Whaling reduced the California stock in the 1850s after the discovery of the breeding lagoons, and again after the turn of the century with the introduction of floating factories. Protection from commercial whaling by international agreement in 1946 has allowed the population to grow to its current estimated level of 15,000 or more animals. Most whaling historians and biologists believe the preexploitation stock size was between 15,000 and 24,000. Alaskan Eskimos from St. Lawrence Island still take a few gray whales annually from skin boats; and Soviet-government catcher boats take about 140 to 200 gray whales each year for use by Siberian aborigines. In 1978 the International Whaling Commission reclassified the gray whale (eastern Pacific stock only) from protected to sustained-management status, and a noncommercial quota is set each year.

CAN BE CONFUSED WITH: In the North Pacific, gray whales are likely to be confused at a distance mainly with right and bowhead whales (the other two large whales without a dorsal fin) and with sperm whales, whose dorsal hump is often small and indistinct. But the well-known coastal migration, the distinctive mottled gray color, and the "knuckles" on the tail stock of the gray whale should make identification easy upon close examination. Some humpbacks have little more than a bump for a dorsal fin, followed by a series of low crenulations along the spine. Humpbacks, however, have many knobs on the head, very long flippers that are usually at least partly white, and ventral pleats or grooves, all characteristics that distinguish them from gray whales.

THE TOOTHED WHALES, DOLPHINS & PORPOISES

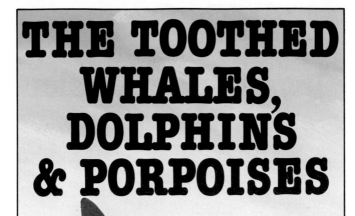

Suborder: Odontoceti

4. The Sperm, Pygmy Sperm, and Dwarf Sperm Whales

Family: Physeteridae

Sperm Whale

Physeter macrocephalus
Linnaeus, 1758
DERIVATION: from the Greek
physeter for "blower"; from the
Greek *makros* for "long," and
kephale for "head."

ALL ZONES

DISTINCTIVE FEATURES: Head huge (to 40 percent of the body length), boxlike; single blowhole on left of head at front; blow projects obliquely forward; body dark grayish brown to brown; skin may appear wrinkled, especially aft of the head; back has rounded or triangular hump followed by knuckles along spine.

DESCRIPTION: Male sperm whales have been reliably reported to reach 18 m in length, though current averages are much smaller, slightly over 15 m. Females are much smaller than males, rarely reaching 12 m in length. Females are sexually mature at eight to eleven years of age, at which time they are 8.3 to 9.1 m long; males at ten or more years of age and 10 to 12 m long. At physical maturity, which is reached slightly later in males than in females, males are about 15.8 m in length and weigh 43.5 tons; females, 10.9 m and 13.5 tons. Newborn are 3.5 to 4.5 m long and weigh about a ton.

The sperm whale is among the easiest of whales to identify at sea, even when comparatively little of the animal is visible. It has a huge head, which makes up from 25 to 33 percent of the animal's total length. The proportion is considerably higher for males than for females, and is the chief cause of the vastly different appearance between adult females and males; in males the spermaceti organ grows to project further beyond the tip of the skull. The blunted squarish snout, which may project up to 1.5 m beyond the tip of the lower jaw, houses a large reservoir containing a high-quality oil called spermaceti, which has long been prized by whalers.

The single blowhole is located well to the left of the midline and far forward on the head. Its generally bushy blow, usually less than 2.4 m long, emerges forward at a sharp angle from the head and toward the left. Under good wind conditions, this feature

An aerial perspective of a pod of sperm whales; 21 individuals can be counted in the photograph. (Off Japan: Suisan Koku Company, courtesy of Toshio Kasuya.)

alone may permit positive identification of sperm whales, even at long distances.

Sperm whales have a distinct dorsal hump, usually rounded or triangular at its peak, located about two-thirds of the way back from the tip of the snout. Immediately behind the hump is a series of knuckles or crenulations along the midline. This hump and the crenulations are clearly visible when the animal arches the tail before beginning a dive. There is a ventral keel, which may also be visible as the animal "sounds" (dives). The flukes of sperm whales are broad and triangular, straight rather than concave on the rear margin, and deeply notched.

Sperm whales usually appear dark brownish gray. The body has a corrugated or shriveled appearance. The belly and front of the head may be grayish to off-white. The skin around the mouth, particularly near the corners, is white. Calves are much lighter gray overall. The narrow underslung lower jaw contains eighteen to twenty-five functional teeth, which are thick and conical and fit into sockets in the generally toothless upper jaw.

NATURAL HISTORY: Sperm whales may be found singly or in groups of fifty or more individuals. Older males usually are solitary or in small groups, except during the breeding season, when they may join with nursery or maternity schools for mating. During the remainder of the year, large groups may be bachelor herds (sexually inactive males), juvenile or immature schools, or nursery schools containing females and juveniles of both sexes. There is currently a debate whether most breeding is accomplished by the seasonally present bulls or by younger bulls, which may remain with a school for most of the year.

Males and females exhibit different migratory behavior. Males range more poleward than females and immature males; females and young are not often found outside the zone bounded by 40° N and 40° S. Most sperm whales of either sex and all age classes shift poleward in spring and summer, returning to temperate and tropical portions of their range in fall.

Sperm whales may dive deeper than 1,000 m (entangled sperm whales have been recorded from that depth, and sonar operators have passively and actively tracked individuals to depths of 2,000 and 2,800 m, respectively), and remain submerged for periods of an hour or more. Like most whales when surfacing from a deep dive, sperm whales emit a single explosive blow, and then, depending on the length of the dive, may remain on the surface for more than an hour and blow more than fifty times before beginning the next dive. Shorter periods of time on the surface and fewer blows are more common. Females may dive and remain on the surface for shorter periods than males. When beginning a deep dive, sperm whales throw their broad triangular flukes, dark on the undersides, high into the air, making the thickened keel visible from some angles.

Sperm whales feed primarily on squid, but may occasionally also take octopus and a variety of fish, including salmon, rockfish, lingcod, and skates, to mention a few. Sperm whales have also been discovered with bizarre objects in their stomachs: rocks, sand, a glass fishing float, deep-sea sponges, crab meat, cut meat of baleen whales, clams, and a human boot, which was no doubt separated from its owner before it was ingested. Many of these are considered evidence that sperm whales sometimes grub for food along the sea bottom.

Ambergris, a peculiar substance that occurs in the lower intestine in lumps weighing up to 100 kg, is formed around squid beaks that remain in the stomach. It was once highly prized as a fixative in the perfume industry, and continues to be valuable despite its widespread replacement by synthetics.

DISTRIBUTION AND CURRENT STATUS: Sperm whales are widely distributed in all oceans of the world, between 60° N and 70° S, avoiding only the polar pack ice in both hemispheres. As we said, patterns of migration are different for males and females of various age groups, the largest males tending to make the most poleward intrusions in summer. In most areas they are seldom found in less than about 180 m of water; otherwise they may be encountered almost anywhere on the high seas.

The distribution of sperm whales tends to be clumped. Areas of concentration have been called "grounds" because of the productive whaling activities that took place on them. Though populations in some are reduced substantially, sperm whales continue to frequent many previously important areas.

In the Pacific, sperm whales range to the deeper waters of the southwestern Bering Sea, as far as St. Lawrence Island and the Soviet coast off Cape Navarin, and south to the ice edge of the Antarctic. Prior to modern whaling, the most important sperming grounds were those near the Hawaiian Islands; in Panama Bay; around the Galapagos; in a variety of locations "on the line," i.e., along the equatorial belt, including the vicinity of oceanic islands; near New Guinea; around the Polynesian Islands (smaller ones); off eastern Australia and New Zealand; and in deep waters seaward of the western South American continental shelf (off Ecuador, Peru, and Chile).

In the Atlantic, sperm whales occur as far north as Davis Strait, Denmark Strait, and the west coast of Norway, at least to 65° N, then south to the ice edge in the southern hemisphere. The most important historical grounds in the northern hemisphere have been: portions of the Grand Bank, including an area just southeast of the southern Grand Bank, from 30° N to 40° N and 35° W to 55° W; off the Carolinas; around the Bahamas; off the west coast of the British Isles; and from the Azores and Madeira across the tropical mid-Atlantic. Those most important in the southern hemisphere have been: off Angola; off the southwestern tip of South Africa; around Tristan da Cunha; and in almost contiguous areas along eastern South America, from Brazil to near the Falkland Islands. Sperm whales are common in the Mediterranean Sea.

In the Indian Ocean sperm whales are distributed widely, and were taken in substantial numbers around its western, northern, and eastern rims, especially near Madagascar; on the Mahe Banks;

in the Arabian Sea; and off India and western Australia. They were taken in lesser numbers in open water and around the more southerly islands.

Since its beginning in 1712, commercial sperm whaling has continued sporadically to the present. A population of more than 2,000,000 whales probably still roamed the seas in the mid-1940s, roughly half in the North Pacific. Catches, which peaked at more than 29,000 per year in the mid-1960s, have substantially reduced most stocks. The IWC, which estimates the current population to total 1.5 million, continued to sanction large annual quotas until 1979, when factory-ship whaling for sperm whales was prohibited. The biggest catches were made by the Japanese and Soviet pelagic fleets, with smaller numbers being taken by shore stations at the

The processing of a sperm whale begins. (Taiji, Japan, February 1980: Howard Hall, Living Ocean Foundation.)

Azores, Madeira, Spain, Brazil, Iceland, Peru, and Japan. These shore stations continue to operate. Sperm-whale meat is not considered palatable for humans in many regions; so sperm whales are used primarily for industrial products, such as oil, spermaceti, and meal. However, sperm-whale meat is eaten in some local areas of Japan.

CAN BE CONFUSED WITH: Because of the unusual head shape and blow, sperm whales are unlikely to be confused with any other species when they can be closely examined. If only the back and flukes are seen, however, sperm whales may somewhat resemble humpback whales. Both species arch the back when beginning a dive, raising the fin or hump, and both throw the flukes. The most distinctive differences between the two species are addressed in the section on humpback whales.

At sea the head of a sperm whale may somewhat resemble those of certain beaked whales (particularly *Berardius* and males of *Hyperoodon*). The latter all have distinct beaks (longer in *Berardius*) and a prominent dorsal fin (more prominent in *Hyperoodon*). In addition, the blowhole of these whales is located well back on the head and not, as in the sperm whale, at the front.

Pygmy Sperm Whale

Kogia breviceps
(de Blainville, 1838)
DERIVATION: *Kogia* is probably
a Latinized form of the English
word "codger," for "a miserly
old fellow"; *Kogia* also has been
attributed to a Turk called
Cogia Effendi, who observed
whales in the Mediterranean;
from the Latin *brevis* for "short,"
and from the New Latin
genitive of *cepitis* for "head."

ZONES 1 TO 5

DISTINCTIVE FEATURES: Head sharklike in shape; underslung, sharklike lower jaw; false gill on side of head; twelve to sixteen pairs of teeth (in lower jaw only) that are thin, incurved, and sharp; confined to temperate and tropical latitudes.

DESCRIPTION: Maximum body length is at least 3.4 m; maximum weight is 408 kg. Adult length is said to be 2.7 to 3.4 m, weight 318 to 408 kg. Females and males are similar in size and appearance. Length at birth is about 1.2 m.

The body is robust, with a short but distinctively shaped head, and a narrow tail stock. The head can be squarish or conical in lateral view, and conical in dorsal view. Its shape changes with body size, becoming more boxlike or rectangular, at times actually bulbous, in larger animals. The head resembles that of the sperm whale, but the underslung lower jaw is proportionately smaller, as is the entire head. The jaw ends well behind the tip of the snout. There is no beak. Because of the position and shape of the mouth, beached specimens are often misidentified as sharks. On each side of the head behind the eye is a crescentic or bracket-shaped mark that is lightly pigmented and set off by a dark line posteriorly. Because of its superficial resemblance to the gill slits of fish, this mark has been called a false gill. The blowhole is on top of the head, but left of the midline, and its shape and orientation appear to differ in individuals.

Flippers are situated below and behind the false gills, well forward on the body. Their length can be up to 14 percent of total body length; they are wide at the base, tapering to a dull point. The dorsal fin is low, strongly falcate, and placed behind the center of the back. Its height is less than 5 percent of the total body length. The tail has a concave trailing edge and a distinct median notch.

Coloration is dark bluish gray dorsally, shading to lighter gray laterally, and gradually fading to a dull white or pink on the belly. The outer margin of the flippers and upper surface of the flukes are steel gray. In some sightings of free-swimming individuals of this species and of dwarf sperm whales, the body has appeared wrinkled, like the body of the sperm whale.

There are no functional teeth in the upper jaw, but twelve to sixteen pairs in the lower (occasionally ten or eleven). The teeth have been described as "thin, curved (inward), and sharp-pointed," and "strongly reminiscent of the teeth of pythons."

NATURAL HISTORY: This whale is very rarely observed at sea, even though the incidence of strandings indicates that it is common, at least seasonally, close to shore in some areas. It apparently is not gregarious, and six or seven animals together would constitute a large group. Most reliable sightings have occurred when the seas were flat and visibility conditions excellent, and have involved from one to five rafting animals.

Females reach sexual maturity at a length of 2.6 to 2.8 m, males at 2.7 to 3 m. Most calving appears to take place between autumn and spring, after a gestation period believed to last about eleven months. Stranded females accompanied by calves have been found to be both pregnant and lactating, suggesting they are capable of annual reproduction.

Hypotheses about seasonal north-south or inshore-offshore movements have been offered; but since these have been based only on strandings, they are highly speculative. In South African waters, pygmy sperm whales appear to be nonmigratory, and there are stranding records for every month of the year in eastern North America.

The few descriptions of behavior at sea allow only tentative generalizations. The whale reportedly rises slowly to the surface to breathe, produces an inconspicuous blow, and does not normally roll aggressively at the surface like many other small cetaceans. At rest it may remain motionless at the surface, with rostrum barely awash and tail hanging down loosely. In this posture the pygmy sperm whale reportedly floats higher in the water, with more of the head and back exposed, than the dwarf sperm whale. If this report is confirmed by future observations, it might prove useful for distinguishing these two species at sea. If startled while in this "basking" posture, this whale may defecate, issuing a cloud of reddish brown feces, then dive out of sight. The pygmy sperm whale is not regarded as a fast swimmer. Cephalopods (squid and octopus) are the dietary staple, but stomatopods, crabs, and fish are also eaten. No predators are known. Individuals or female-calf pairs often strand alive. Many have been rushed from beaches to aquariums, only to die within a few days or weeks.

DISTRIBUTION AND CURRENT STATUS: The pygmy sperm whale appears to be a cosmopolitan species, recorded from nearly all temperate, subtropical, and tropical waters. Knowledge of its distribution is based almost exclusively on stranding records, and even this is clouded by failures to distinguish it from the dwarf sperm whale. Preliminary analyses of stomach contents indicate that *K. breviceps* is principally oceanic in distribution, staying most often seaward of the continental shelf.

Strandings are most common along: the east coast of North

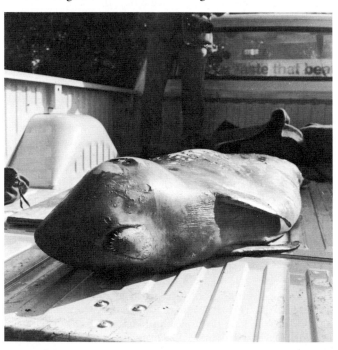

Much of what is known about pygmy and dwarf sperm whales comes from stranded specimens. (Corolla, North Carolina: Barry Peers, courtesy of James G. Mead.)

America, from Nova Scotia (Sable Island and Halifax) south to Cuba and Texas; southern Africa in the vicinity of Cape of Good Hope; southeastern Australia; and the Tasman Sea and South Pacific coasts of New Zealand. However, strandings have also confirmed its presence off Peru; the west coast of North America, from Washington State south to (and into) the Gulf of California; Japan; India; East Africa; and western Europe (Netherlands and France).

Nothing is known about abundance or stock distinctions. However, the pygmy sperm whale is said to be the second most common stranded cetacean (after the bottlenose dolphin) in the southeastern United States. Some casual exploitation occurred at the hands of early whalers, and a few have been (and may still be) taken with hand harpoons off southern Japan and Indonesia.

CAN BE CONFUSED WITH: The pygmy sperm whale is regularly confused with the dwarf sperm whale, which is smaller (2.7 m maximum length), has a taller and more dolphinlike dorsal fin, has fewer teeth (seven to twelve, rarely thirteen, pairs), and has several irregular creases on the throat. The teeth of *K. breviceps* are longer and thicker than those of *K. simus;* any *Kogia* specimen with teeth longer than about 30 mm and more than 4.5 mm in diameter is almost certainly *K. breviceps*. At sea the pygmy sperm whale's dorsal fin might allow it to be confused at a distance with small beaked whales of several species. However, on closer inspection the distinct beak of beaked whales should allow easy differentiation.

Dwarf Sperm Whale

Kogia simus
Owen, 1866
DERIVATION: from the Latin
simus for "flat-nosed."

ZONES 1 TO 5

DISTINCTIVE FEATURES: Similar to *K. breviceps*, but smaller, and with one to three pairs of maxillary teeth (sometimes present), several irregular gular grooves (occasionally present), and a larger, more erect dorsal fin, positioned farther forward on the back; other differences rather slight; confined to temperate and tropical latitudes.

DESCRIPTION: Maximum size is 2.7 m and 272 kg, with adults ranging from 2.1 to 2.7 m and 136 to 272 kg. Length at birth is about 1 m.
   Generally this species is similar to *K. breviceps*, with the exceptions noted. The dwarf sperm whale's snout is consistently shorter (more pugged) than the pygmy sperm whale's, but must be measured carefully to be of use for identification. The shape of the head changes markedly with age. In young individuals the dorsal margin of the snout slopes downward anteriorly; in older animals it is straight, giving a squarish profile in lateral view. The dwarf sperm whale's dorsal fin is tall and falcate, closely resembling that of the bottlenose dolphin, and it is normally located near the center of the back. The height of the dorsal fin is more than 5 percent of the total

body length, and the base of the fin is generally somewhat longer in *K. simus* than in *K. breviceps*. There can be several short, irregular creases or grooves on the throat of the dwarf sperm whale, similar to those found on adult sperm whales. Like pygmy sperm whales, dwarf sperm whales may appear wrinkled, a feature reminiscent of the sperm whale.

This species has seven to twelve, rarely thirteen, pairs of mandibular teeth. In contrast to the pygmy sperm whale, it occasionally has up to three pairs of rudimentary maxillary teeth as well. The teeth of *K. simus* are shorter and proportionately more slender than those of *K. breviceps;* any taken from a *Kogia* and found to be longer than 30 mm and more than 4.5 mm in diameter can be assumed to be from *K. breviceps*.

NATURAL HISTORY: Like the pygmy sperm whale, the dwarf sperm whale is usually encountered in groups of no more than ten animals. There may be three kinds of pods: one consisting of females with calves, one of immatures, and another of adults of both sexes unaccompanied by calves.

Both males and females apparently reach sexual maturity at a length of 2.1 to 2.2 m. The calving season appears to be prolonged, covering at least five to six months. Pregnant females accompanied by unweaned calves have been observed often enough to suggest that there may be an annual reproductive cycle.

The diet of dwarf sperm whales consists primarily of squid, although fish and crustaceans are also eaten. Stomach contents indicate that the species can dive to depths of at least 300 m. Seasonal migrations for this species have not been documented.

DISTRIBUTION AND CURRENT STATUS: Since the dwarf sperm whale was recognized only recently as a distinct species, and even more recently given its own common name, most historical records of it have been lumped in with those of the pygmy sperm whale. As a result, the limits and extent of its range are not well-defined. Along the east coast of North America, it is known from Virginia south to the Lesser Antilles and throughout the eastern and northern Gulf of Mexico; on the west coast of North America, it has been recorded only from central California and the Pacific coast of southern Baja California. It has also been recorded off South Africa, India, Sri Lanka, Japan, Hawaii, Guam, and South Australia.

The stomach contents of dwarf sperm whales stranded in South Africa indicate that the species' distribution may be somewhat more inshore than that of the pygmy sperm whale, perhaps centering along the edge of the continental shelf.

Like the pygmy sperm whale, this species was probably killed opportunistically by early whalers, and it has been taken by coastal fishermen off southern Japan, Indonesia, and the Lesser Antilles.

CAN BE CONFUSED WITH: Because of its tall, falcate dorsal fin, the dwarf sperm whale may be confused at a distance with any of the small dolphin species. With the pygmy killer whale and melon-headed whale, confusion is likely because of the dwarf sperm whale's dark steel-gray coloration and blunted head. The former two species, however, assemble into much larger aggregations; also, they swim faster and with much more animation. A good look at the head is necessary to distinguish these three species.

Currently available information does not offer a reliable way to distinguish dwarf and pygmy sperm whales in sightings at sea, although the size and placement of the dorsal fin, and the corresponding length of the back that is visible, may prove to be consistently different.

5. The Narwhal and the White Whale

Family: Monodontidae

Narwhal

Monodon monoceros
Linnaeus, 1758
DERIVATION: from the Greek
monos for "one, single," *odon* for
"tooth," and *keros* for "horn."

ZONES 7 AND 8

DISTINCTIVE FEATURES: No dorsal fin; head rounded with slight hint of beak; superficially toothless, except for spiraled tusk of males (and occasionally of females); trailing edge of flukes strongly convex in adults; mottled coloration on dorsum of adults; old animals nearly all white; circumpolar in Arctic above 65° N.

DESCRIPTION: Maximum body length is approximately 5 m (exclusive of the tusk); weight 1,600 kg. Males are larger than females. Newborn calves are 1.5 to 1.7 m long and weigh about 80 kg.

The head is small and rounded. The bulbous forehead so dominates the head profile that the slight beak is insignificant. The crescentic blowhole is to the left of center on top of the head.

The relatively short flippers have noticeably upturned tips, a feature that becomes more exaggerated with age. There is no dorsal fin, but an uneven ridge about 5 cm high runs along the spine in the half nearest the tail. The flukes are fan-shaped, with a deep notch and slightly rounded tips. The trailing margin of the flukes becomes more strongly convex with age.

Skin color also alters with age. Newborn are blotchy slate gray or bluish gray. Juveniles become completely bluish black or black. As they mature, white streaks and patches appear around the anus, genital slit, and navel, spreading over the entire ventral surface and onto the flanks. Old animals, especially males, can be almost entirely white, with flecks of dark pigmentation only on the head and neck.

The narwhal's small, narrow mouth is toothless. Adult teeth, of which there are only two, remain firmly embedded in the upper jaw. Normally in males the left tooth protrudes through the gum, directed forward and with a left-hand spiral. Occasionally both teeth are extruded, the right one usually being imperfectly developed. External tusks in females are exceptional. Narwhal tusks can grow to be more than 3 m long, with a basal circumference of more than 20 cm. Large tusks can weigh up to 10 kg.

NATURAL HISTORY: Mass movements of thousands have been observed, but such large herds are subdivided into smaller, more cohesive units of twenty or fewer individuals. Bands of juveniles,

The adult male narwhal's spiraled tusk is one of the most remarkable appendages in the animal kingdom. (Eastern Canadian Arctic: Kerwin J. Finley.)

"nursery" groups of females and calves, and aggregations of tusked bulls have been seen, but mixed groupings are not unusual. Females with calves are more accessible to hunters in some areas, since large males seem to remain farther from shore. Narwhals are sometimes seen in the close company of white whales.

Sexual maturity is probably reached at about four to seven years in females, eight to nine years in males. The breeding season is thought to be April. Calves are born in summer after a gestation period of about fourteen to fifteen months. Lactation probably lasts a year or more. The interval between calves may average three years.

This species is not often found far from loose pack ice. Its inshore movement in summer is dramatic and predictable, and coincides with the breakup of ice cover. Annual migrations appear to be responsive to ice formation and drift. Entrapment by fast-forming ice in the fall is not uncommon, and may contribute significantly to natural mortality.

The purpose of the narwhal's tusk has baffled scientists until recently. At various times it has been described as a spear for catching large fish, an ice pick, and a hoe for grubbing bottom sediments in search of food. Although some early authors speculated about its suitability as a weapon, not until very recently did some observations support the idea that the tusk's primary function is aggressive. Scarring on the head of adult males is now thought to be

A group of eight tusked male narwhals passes beneath a survey aircraft in the eastern Arctic of Canada. (Wybrand Hoek.)

caused by tusks of competitors; they probably engage in some kind of jousting or sparring during the mating season, in spring or late winter. Many old males have broken tusks, and the distal portion of unbroken tusks is smoothly polished in contrast to the algae-coated proximal portion.

A narwhal will sometimes "stand" vertically in the water, with its head (and tusk if present) exposed. Narwhals do not normally breach, but they sometimes lunge as they surface, exposing the head and much of the forward half of the body. The tusk is occasionally, but certainly not always, visible as the animal rolls at the surface.

Known prey includes squid, polar cod, demersal fish, and crustaceans. Narwhals probably are preyed on by killer whales, rarely by polar bears, possibly by walruses. Greenland sharks consume narwhal carcasses, but probably attack only weak or badly wounded individuals.

DISTRIBUTION AND CURRENT STATUS: The narwhal is locally abundant in the high Arctic, especially in the western hemisphere. Large concentrations are known to occur in Davis Strait, in Baffin Bay and adjacent waters, and in the Greenland Sea. Smaller numbers inhabit Hudson Strait, northern Hudson Bay, Foxe Basin, and the Barents Sea. The narwhal's presence in the Beaufort, Bering, and eastern Chukchi Seas is exceptional. This species penetrates deep into the permanent polar pack in summer, with the birth of a calf recorded north of 84° N.

Several thousand are said to remain in the Soviet Arctic, possibly divided between an eastern and a western stock. The only population for which there is a reliable estimate of minimum abundance (20,000 or more) is the stock that enters Lancaster Sound in summer.

The narwhal was traditionally hunted by northern natives for subsistence, and the commercial value of its tusk made it attractive to European and American whalers and traders. Today it is hunted in Greenland and Canada for its ivory, and to a limited extent for its meat and skin.

Narwhals can be seen during August near the settlements of Pond Inlet and Arctic Bay in the eastern Canadian Arctic, and throughout the summer in Inglefield Bay in northwestern Greenland.

CAN BE CONFUSED WITH: Its close relative, the white whale or beluga, is the only cetacean with which the narwhal is likely to be confused. The tusk, if present, is a distinguishing feature, and the dark splotches on the head and back of the adult narwhal contrast with the even whiteness of the mature beluga. Confusion is more likely for very young or very old animals. The most recent alleged sighting of the extinct Steller sea cow, at Cape Navarin in the western Bering Sea, was possibly of a female narwhal.

White Whale
or Beluga

Delphinapterus leucas
(Pallas, 1776)
DERIVATION: from the Greek
delphinos for "dolphin," *a* for
"without," *pteron* for "fin," and
leukos for "white."

ZONES 2, 7, AND 8

DISTINCTIVE FEATURES: Adult coloration completely white; no dorsal fin; short, broad beak; broad, spatulate flippers; circumpolar in Arctic, also found in subarctic areas.

DESCRIPTION: White whales reach a maximum length of about 4.5 m and a maximum weight of 1,500 kg, although there are marked size differences between geographically separate populations. Males are somewhat larger than females. Newborn are about 1.5 m long.

White whales have robust bodies that taper at both ends, with a proportionately small head and a narrow caudal peduncle. There is a short but distinct beak, which is often overhung by a prominent, rounded melon. The melon readily changes shape, and the white whale's free cervical vertebrae allow it to nod and turn its head as few other whales can. The blowhole is a transverse slit located just ahead of the neck crease.

The flippers are short, broad, and spatulate in shape. They are upcurled at the tips in adults. Instead of a dorsal fin, the white whale has a narrow ridge along the spine just behind the midpoint of the back. This ridge may be notched laterally to form a series of small bumps, and it is often darkly pigmented. The tail is broad and ornately curved on the trailing edge, whose convexity becomes more exaggerated with age. There is a median notch between the flukes.

At birth belugas are slate gray to pinkish brown. As they grow older, they become a uniform blue or bluish gray. Sometime after attainment of sexual maturity, they become completely white.

There are eight to eleven teeth in each upper jaw; eight to nine in each lower jaw. These teeth are irregular in form, are often curved, and sometimes show considerable wear.

NATURAL HISTORY: This species is gregarious, and can be seen in aggregations of more than a thousand animals, particularly in estuaries where large concentrations occur in summer. Mating occurs in spring, and young are born in summer after a gestation period of about fourteen months. Calves remain with their mothers for two years, though they eat some solid foods during the second year. Females are believed to become fertile at about five years of age, males at eight to nine. The calving interval is two to three years.

Some populations are strongly migratory, others basically resident in a well-defined area. The primary determinant of white-whale movements is ice cover. It is not known why white whales are attracted in summer to relatively warm estuaries and to fresh

A pod of belugas follows a lead through the pack ice, heading toward their estuarine summer grounds. (Northern Chukchi Sea, May 28, 1980: Sue Moore.)

The predicament faced by white whales that misjudge water depth and strand as the tide changes need not be fatal. If unmolested, they often escape to deeper water during the next flood tide. (Cunningham Inlet, Somerset Island, eastern Canadian Arctic, July 1977: Wybrand Hoek.)

waters far up rivers. The best-known and perhaps most impressive long-distance migration of this species is undertaken by a population believed to winter in the Bering Sea. These whales pass through Bering Strait, and follow the spring lead (often in company with bowhead whales) along the northern Alaskan and Yukon coasts. They congregate by the thousands in the Mackenzie River delta of the eastern Beaufort Sea during the summer. In September the buildup of ice forces them westward once again.

Because it is often found near shore, the white whale is usually called a shallow-water, coastal species. However, it is also found in deep water, and it may dive as well as other deep-water cetaceans. The white whale is not as demonstrative as some other cetaceans; it almost never leaps clear of the water (except after training in captivity). However, the white whale's chirps and squeals create a great noise underwater, and can often be heard in air as well. These whales adapt well to captivity; some have lived more than a decade in confinement.

Natural predation on white whales is not thought to be heavy, although killer whales, polar bears, and possibly an occasional rogue walrus hunt them. Ice entrapment and consequent starvation, suffocation, or predation by man or bears may take a significant toll. White whales are believed to live at least 25 years.

DISTRIBUTION AND CURRENT STATUS: The white whale is confined to arctic and subarctic waters. Because it was long hunted both commercially and for subsistence, it is relatively well-studied in most parts of its range. Areas where white whales are known to range (with recent population estimates in parentheses) are as follows: Cook Inlet, Alaska (300); Bristol Bay (1,000–1,500); Beaufort Sea (4,500; may include some animals from Bristol Bay); Lancaster Sound and adjacent areas (10,000); throughout Baffin Bay and along the west side of Davis Strait (several thousand); western Hudson Bay and Hudson Strait (more than 11,000); Gulf of St. Lawrence (500); Greenland Sea; Barents Sea; White Sea; Kara Sea; Laptev Sea; East Siberian Sea; Chukchi Sea; Bering Sea; and Sea of Okhotsk. Those populations inhabiting the Gulf of St. Lawrence, Cumberland Sound in western Davis Strait, Ungava Bay in Hudson Strait, and eastern Hudson Bay have been severely depleted by historical overexploitation and other forms of disturbance.

Local fisheries that use the technique of driving white whales ashore have been prosecuted in many areas, including western Hudson Bay, Prince Regent Inlet, Cumberland Sound, and the Gulf of St. Lawrence. Stationary and trawl nets have been used extensively for harvesting white whales in the Soviet Arctic. Native whalers in Alaska, Canada, and Greenland hunt them with rifles and harpoons. Oil, muktuk, and meat are the principal products, although "porpoise leather" bootlaces were manufactured from white-whale skins in England during the early twentieth century.

Commercial hunting has ended (except in the Soviet Union), but native hunting continues to put pressure on many stocks. Habitat alteration, especially the damming of rivers on whose outflow white whales might depend for their summer assembly sites, may be a more serious problem than hunting. Artificial island building, pipeline construction, and tanker and icebreaker traffic in polar regions pose potential threats to white whales.

The most accessible area for observing this species is near Tadoussac, Quebec, where the scenic Saguenay River, populated by white whales during summer and fall, empties into the St. Lawrence.

CAN BE CONFUSED WITH: The species with which the white whale is most likely to be confused is the narwhal. The two are occasionally seen together. Only neonates and old individuals are likely to present problems, however. Newborn narwhals have a blotchy, irregular coloration, but young white whales are an even brown or blue gray. The best way to tell them apart is probably by the appearance of the accompanying parent. Narwhals that are nearly pure white are generally old bulls, and since these almost invariably have large, conspicuous tusks, there should be little confusion.

Individual white whales occasionally wander into cold temperate regions, especially in the Atlantic, where they can be confused with Risso's dolphins and Cuvier's beaked whales. The last two species have a dorsal fin, however: prominent and darkly pigmented in Risso's dolphin; more modest and positioned well back on the body in Cuvier's beaked whale. Also, only large adults of these two species are likely to be white enough to look like belugas, but they will have some darker pigmentation on the appendages.

6. The Beaked Whales

Family: Ziphiidae

Baird's Beaked Whale

Berardius bairdii
Stejneger, 1883
DERIVATION: Bérard was the
French commander of the
vessel that carried the type-
specimen of the genus from
New Zealand to France in 1846;
Spencer Fullerton Baird
(1823–1887) was a celebrated
American naturalist and
Secretary of the Smithsonian
Institution.

Arnoux's Beaked Whale

Berardius arnuxii
Duvernoy, 1851
DERIVATION: Arnoux, a French
surgeon, was aboard Bérard's
vessel that carried the type
specimen from New Zealand to
France.

ZONES 1 AND 7
ZONES 3 TO 6

BAIRD'S BEAKED WHALE

DISTINCTIVE FEATURES: Bulbous, steep forehead (melon); dolphinlike snout; numerous scratch marks on back; posterior positioning of triangular or falcate dorsal fin; two pairs of laterally compressed teeth at tip of lower jaw; lower jaw extends beyond upper, exposing apical mandibular teeth in adults; *bairdii* limited to North Pacifici, *arnuxii* to southern oceans.

DESCRIPTION: The case for recognizing two rather than one species in the genus *Berardius* is by no means conclusive, though the two forms appear to be geographically separate. The southern form is thought to be smaller than the northern form, and there are osteological differences between the two. Very few Arnoux's beaked whales have been examined and described, so most of this account, except where noted, applies to the better-known Baird's beaked whale.

Physical maturity is usually attained at about 11 m in females and 10.7 m in males, though maximum lengths are said to be about 12.8 m in females and 11.8 m in males. Females become sexually mature at 10 to 10.3 m; males, 9.3 to 9.6 m. At birth these whales measure about 4.5 to 4.6 m. Arnoux's beaked whales are said to attain a maximum length of about 9 m, but the sample of measured specimens is small. In this species sexual dimorphism in size has not been demonstrated.

Both sexes have a prominent, bulbous forehead that slopes smoothly, but steeply, to a long, tubelike beak. This beak can often be seen when an animal brings its head out of the water at a steep angle as it surfaces to breathe. At the front of the melon there is a visible concavity. The crescentic blowhole is at the center of the top of the head, with the rounded side facing anteriorly. This orientation of the blowhole is a characteristic unique to the genus *Berardius*. The top of the lower jaw extends slightly beyond that of the upper jaw. Occasionally short, irregular furrows are present between the two long (up to about 70 cm), V-shaped throat grooves.

The body is elongated and rotund. The flippers do not taper, and are rounded distally. A proportionately small, triangular dorsal

A stranded specimen of the rarely observed Arnoux's beaked whale. (Hawkes Bay, New Zealand, November 1968: Frank Robson, courtesy of Kenneth C. Balcomb.)

fin is situated more than two-thirds of the way toward the flukes. This fin is rounded slightly at the top, and not usually falcate in *B. bairdii,* though it is at least sometimes small and distinctly falcate in *B. arnuxii.* The flukes, which sometimes are raised as the whale dives, have an almost straight or concave trailing edge, with only a shallow, indistinct concavity or a slight convexity near the midline. There is usually no median notch between the flukes.

The color of both species ranges from slate gray to army brown to black, with white blotches on the underparts, particularly on the throat, between the flippers, and at the umbilicus. Arnoux's beaked whales generally are light gray to white ventrally, with occasional pale patches on the back. They also have many circular white scars on the belly and sides. Males and, to a lesser extent, females in both species have many linear white scars on the body, apparently because of intraspecific fighting. These often make the dorsal surface appear lighter than it otherwise would.

Males and females have a pair of triangular, laterally compressed teeth that erupt near the tip of the mandibles in both sexes; a second, smaller pair is present behind these. The anterior pair erupts when the animal is at least several years old, and the rear pair erupts later in life. These species (together with the Tasman beaked whale) are exceptional among beaked whales, in that females have functional teeth. The apical teeth are exposed outside the mouth and flash brightly in sunlight, making them especially noticeable from the air. The teeth of old individuals may be so worn that they are even with the gum.

NATURAL HISTORY: Since catches off Japan and British Columbia have consisted mainly of males, there may be sexual segregation, at least seasonally, within the populations of Baird's beaked whales. Groups of two to twenty animals are usually seen swimming in close formation; sometimes as many as thirty are seen together. Most records of *B. arnuxii* are of stranded individuals; so knowledge of the species' natural history is very limited.

Sexual maturity in *B. bairdii* is reached at an estimated age of eight to ten years. Gestation probably lasts at least ten months, and may be as long as seventeen months. Mating generally occurs in October and November, and most births have been recorded from November to July, with a peak in March and April. The calving interval is assumed to be about three years.

Seasonal movements are not well-understood, although Baird's beaked whale is seen off northern California from at least June to October, and off British Columbia between May and September. For the population near Japan, seasonal north-south movements have been postulated.

The majority of strandings of *B. arnuxii* around New Zealand have taken place in spring or early summer (December to February or March), and ice-entrapped whales were observed off Graham

The bulbous melon and long beak of a Baird's beaked whale. (Off northern California, 40° 19' N, 124° 42' W, June 26, 1980: Gary L. Friedrichsen.)

Land in the Antarctic in April. However, seasonal shifts in distribution for this species have not been confirmed.

Baird's beaked whales are often difficult to approach. Their blow is low and indistinct. They roll sharply when swimming, and sometimes show their flukes as they dive. Several shallow dives are frequently followed by a deep dive that may last as long as twenty minutes. The spout is low and bushy.

The diet of these whales consists of deep-sea fish, octopus, and squid. They also are known to eat rockfish, mackerel, sardines, crustaceans, and sea cucumbers in the North Pacific.

Natural mortality factors are unknown, but ice entrapment may be significant in the southern hemisphere. An Arnoux's beaked whale trapped in ice remained alive for about six months before it was shot. It was seen to jump almost clear of the water four times within a minute and a half. These whales are believed to live for seventy years or longer.

DISTRIBUTION AND CURRENT STATUS: Baird's beaked whale is confined to the North Pacific, and it is usually found only far offshore, in waters deeper than 1,000 m. On the eastern side it has been observed from the Pribilof Islands and Alaska south to Baja California, and on the western side from the Kuriles, Kamchatka, and the southern Sea of Okhotsk to southeastern Japan, generally in pelagic waters.

A few were killed in the defunct coastal whale fisheries off California and British Columbia, and by pelagic whalers from the U.S.S.R. The only sizeable fishery was located in Japan, where catches of several hundred per year were made in the 1950s. Recent hunting of Baird's beaked whale is limited to coastal waters of the

Once hunted off California and northern Hokkaido as well, Baird's beaked whales are now taken only off the Boso Peninsula, Japan. (Pacific coast of Japan: courtesy of Toshio Kasuya.)

Boso Peninsula on the Pacific coast of Japan, where maritime traffic to and from Tokyo Bay may constitute a significant disturbance.

There are no estimates of the size of any population of Baird's beaked whale. Heavy catches during the 1950s may have reduced the stocks near Japan.

Arnoux's beaked whale has been reported only in the South Pacific and South Atlantic oceans, between 30° S and Antarctica. It almost certainly inhabits the southern Indian Ocean as well. The relative frequency of reported strandings indicates that the waters around New Zealand may be an area of concentration. Other known areas in which they are at least sporadically present include the waters near South Australia, Argentina, the Falkland Islands, South Georgia, South Shetlands, South Africa, and Graham Land (Antarctic Peninsula). Recorded sightings of free-swimming animals are all from pelagic waters, but were often near seamounts or other significantly steep bottom slopes.

CAN BE CONFUSED WITH: It is difficult to distinguish these species from bottlenose whales (*Hyperoodon* spp.) at sea. However, on close inspection, the relatively longer beak, less bulbous forehead, and smaller dorsal fin of the *Berardius* species may help identify them. Also, the latter have two pairs of teeth in the lower jaw, but bottlenose whales usually have only one pair. The teeth of *Berardius* are more massive and conspicuous than those of *Hyperoodon;* if the teeth are visible from any distance at all, the specimen in question is probably *Berardius*.

There is an equatorial population of bottlenose whales, probably *Hyperoodon planifrons,* in the Pacific. This population is not known to overlap in range with either Baird's or Arnoux's beaked whale, but sightings of large beaked whales with bluff foreheads in tropical Pacific waters should be checked out carefully to ensure against misidentification.

Individuals less than 7 m long might be confused with various species of *Mesoplodon* and with Cuvier's beaked whale. However, the mesoplodonts have slimmer heads and much less bulbous foreheads; Cuvier's beaked whale has a shorter beak and a nonbulbous forehead.

In the southern hemisphere one might confuse Arnoux's beaked whale with the Tasman beaked whale. Also, it is worth noting that young Arnoux's beaked whales are sufficiently similar to *Mesoplodon hectori* to have led one investigator to reject the latter as a valid species, arguing that specimens referred to it were actually young of *Berardius arnuxii.*

At a distance, there is some possibility of confusion with the minke whale, especially if only the back and dorsal fin are seen. However, there is very little resemblance, even superficially, between the two *Berardius* species and the minke whale.

Northern Bottlenose Whale

Hyperoodon ampullatus
(Forster, 1770)
DERIVATION: from the Greek *hyper* for "above" or *hyperoon* for "upper story or room" (referring to the palate), and *odon* for "tooth"; from the Latin *ampulla* for "flask, bottle," and *atus*, a suffix meaning "provided with"; the generic name refers to rough papillae on the palate, which were mistaken for teeth.

Southern Bottlenose Whale

Hyperoodon planifrons
Flower, 1882
DERIVATION: from the Latin *planus* for "level," and *frans* for "brow" or *frons* for "front."

ZONES 2 AND 8
ZONE 6, AND SOUTHERN
ENDS OF 3 TO 5

NORTHERN BOTTLENOSE WHALE

DISTINCTIVE FEATURES: Dolphinlike, bottle-nosed beak, but animal is much larger than any beaked dolphin; extremely bulbous forehead (melon), especially in adult males; usually one pair of teeth at tip of lower jaw, erupted in older males and concealed in females; *H. ampullatus* found only in temperate to arctic North Atlantic; *planifrons* circumpolar in southern hemisphere, and possibly reaching tropical waters of the Pacific and Indian oceans.

DESCRIPTION: Northern bottlenose whales reach lengths of up to

9.8 m and can weigh several tons. They are robust, with recorded girths of as much as 6 m. Males are much larger than females. Length at birth is 3.5 to 3.6 m. The largest measured female southern bottlenose whale was 7.45 m long, male, 6.94 m, but these were probably less than maximum size, if one can judge by the greater dimensions of northern bottlenose whales*.

The bottlenose whales are characterized by a bulbous forehead, which is more pronounced in larger specimens and most distinctive in adult males, and by the dolphinlike beak, which is sometimes visible as the animal surfaces steeply to breathe. Although the lower jaw extends slightly farther forward than the upper, this feature is much less pronounced than in *Berardius*. The single, crescentic blowhole is located in an indented area behind the bulging forehead, and its bushy blow projects slightly forward and to a height of 2 m. The concave side of the blowhole is oriented anteriorly.

The hooked dorsal fin is at least 30 cm high, and is located about two-thirds of the distance back from snout to tail. The flippers are short and tapered distally. The broad flukes have a deeply concave, but unnotched, trailing edge. A V-shaped pair of short grooves, similar to those found on other beaked whales, is found on the throat, with the wide end of the V oriented toward the tail.

Young northern bottlenose whales are a uniform chocolate brown. As they age, they remain chocolate brown on the back, but are often lighter on the sides and belly, frequently with irregular patches or blotches of grayish white on the back and sides. Individuals in a group of bottlenose whales seen near the equator in the Pacific were described by the observer as flaxen on the thorax. They had dark necklines demarcating the light head from the dark back. Some animals are buff or gray rather than brown.

Southern bottlenose whales have been described as cloud gray or bluish black, with increasing paleness toward the flanks; and at least some individuals have small white spots on the ventrum and sides. Most larger individuals probably are extensively scarred. Extremely large specimens, especially older males, have a white head. The flippers and undersides of the flukes are brown or gray.

Bottlenose whales have one pair of fully developed conical teeth in the lower jaw, but these usually erupt through the gum only in older adult males. They are situated at the tip of the mandibles. In males the teeth are much stouter than in females, and they show peculiar wear. A second pair of developed teeth is sometimes found during museum preparation just behind the first; some individuals have a series of vestigial teeth the size of toothpicks in the upper and/or lower jaws.

---

* The northern species is much better known than the southern, and the description and data in this account are based on observations made in the North Atlantic, except as noted.

NATURAL HISTORY: These whales appear to have strong social ties; they are often found in groups of four to ten, occasionally more, but pairs and solitary individuals are also commonly seen. There is some evidence of sexual segregation during migration in the North Atlantic, although a pod observed near the equator in the Pacific seemed to include various age classes and both sexes. It has been suggested that large, dominant males may establish and maintain the integrity of groups of females and young. Bottlenose whales in Pacific equatorial waters have been seen in the company of pilot whales.

Minimum length at sexual maturity for northern bottlenose whales is 6.7 m for females and 7.3 m for males, at about seven years for both sexes. Mean length at sexual maturity has been estimated as about 6.9 to 7.5 m, or at ages of eleven years for females, and seven to eleven years for males. The peak season for bottlenose-whale breeding in the North Atlantic is spring (especially April), and since gestation lasts about twelve months, this is also the time of most parturition. Calves probably are not weaned for a year or more, and the interval between pregnancies is two to three years.

The northern whales follow a relatively well-defined migratory pattern. They are found at low latitudes only during winter, but by early spring (March and April) are already present in subarctic regions. They summer in subarctic and arctic latitudes, and begin to drift southward in late summer and early fall. Some bottlenose whales almost certainly winter in high latitudes, perhaps even along the margin of pack ice. Nothing is known about seasonal movements in the southern hemisphere, but a small group of southern bottlenose whales was seen in the Antarctic in summer (January), and bottlenose whales have been seen in the tropical Pacific in February and August.

Bottlenose whales are deep divers; in fact, they seem able to submerge for much more than an hour. They are seldom found in waters shallower than 100 fathoms (183 m). After a long dive, they usually remain at the surface for ten minutes or more, blowing at regular intervals before making another dive. After the last blow in a series, or when startled by a boat, a whale will sometimes show its flukes as it begins to dive. In the North Atlantic, bottlenose whales have been seen lobtailing and breaching.

Two well-developed behavioral traits have made the northern bottlenose whale especially vulnerable to overhunting. First, it approaches vessels, and will stay near a drifting or idling craft for a long time. Second, members of a pod will not desert a wounded companion until it is dead. This care-giving tendency allows whalers to destroy a large proportion of a given pod before it disperses.

The staple of this whale's diet is squid (*Gonatus fabricii*), although it also feeds on pelagic fish (e.g., herring) and even on sea stars, at least occasionally.

No natural predators, except possibly killer whales, are known,

Nineteenth-century Scottish bottlenose whalers called old bulls "flatheads," referring to the squarish melon. This specimen was taken by Norwegian whalers during a recent episode of pelagic whaling. (North Atlantic: Ivar Christensen.)

and bottlenose whales probably live at least 37 years. Strandings occasionally occur; at least one stranding of four animals in the North Atlantic has been recorded.

DISTRIBUTION AND STATUS: Northern bottlenose whales are found in cold temperate to arctic waters of the North Atlantic, often along boundaries between cold polar currents and warmer Atlantic currents. They appear to prefer deep water, 1,000 m or more in depth. Although they appear to prefer the ice edge, it is not unusual for bottlenose whales to be found in broken pack ice.

In the western North Atlantic, Rhode Island is the southern

limit of their known range. The Gully southeast of Sable Island and the northern Labrador Sea near the entrance to Hudson Strait are areas of known concentration, and bottlenose whales are probably year-round residents of these two areas. In spring and summer they spread northward in Davis Strait to the edge of pack ice, sometimes visiting deep channels of the Gulf of St. Lawrence, and bays or fjords off eastern Newfoundland.

On the eastern side of the North Atlantic, they winter along the Atlantic coasts of western Europe, possibly including the Baltic and Mediterranean Seas. Stranding and catch records suggest that they pass Dutch and British coasts on their way north during spring, and again on their way south in late summer and early fall. They summer primarily in the Norwegian and Greenland Seas, near Iceland, Jan Mayen, and Spitsbergen; and, in smaller numbers, in the Barents and White Seas.

Intensive fisheries were conducted beginning in the late nineteenth century by Norway and (much less intensely) by Scotland and later Canada. Norwegian sealers killed more than a thousand in the Greenland Sea in 1885, and an average of 2,500 per year during the 1890s. Although the fishery ceased in the 1920s, it resumed after the Second World War and continued into the late 1960s. As a result, the stocks in the eastern North Atlantic near the Faeroes, Iceland, Jan Mayen, Svalbard (Spitsbergen and Bear Island), and Norway may be depleted. Oil and animal food were the main products. These are of little value at present, and the species is not now being hunted on a major scale.

The southern-hemisphere species (*H. planifrons*) has a circumpolar distribution, with records from Brazil, Argentina, the subantarctic islands of the South Atlantic, Chile, New Zealand, Australia, South Africa, and the Pacific and Indian Ocean sectors of the Antarctic. Its northern limit was thought to be 20° S, but recent evidence indicates that it may range into equatorial waters of the Pacific. The bottlenose whales seen near the equator on several occasions have not been identified with any certainty, but some investigators believe them to be of the southern form. The southern bottlenose whale is rarely observed, and has never been hunted commercially on a major scale. A few have been taken by Soviet catcher boats associated with pelagic floating factories.

CAN BE CONFUSED WITH: Within its range the northern bottlenose whale may be confused with Cuvier's beaked whale, Sowerby's beaked whale, and possibly minke and sperm whales. The forehead of Cuvier's beaked whale is less bulbous or steep, and its beak shorter and less well-defined; also its distribution is generally more tropical in the North Atlantic than that of the northern bottlenose whale. Sowerby's beaked whale certainly overlaps in distribution with the northern bottlenose whale, but its much smaller size and less bulbous melon should help distinguish it. Minke whales should be easy to recognize at close range by their head shape, shiny black skin, and white flipper patch. The sperm whale's angular blow, wrinkled skin, deeply notched flukes, and lack of a beak should make it fairly easy to distinguish.

The southern bottlenose whale can easily be confused with Arnoux's beaked whale at sea. Slight differences in dorsal fin size (small in *Berardius*) and prominence of forehead (greater in *Hyperoodon*) might help. In dead animals, dentition can distinguish the two: bottlenose whales usually have a single pair of exposed teeth (males only); Arnoux's beaked whales have two pairs. Southern bottlenose whales, especially females and young, might be mistaken for species of *Mesoplodon* and Cuvier's beaked whale, as well as the minke whale and sperm whale.

Cuvier's Beaked Whale

Ziphius cavirostris
G. Cuvier, 1823
DERIVATION: probably from the Greek *xiphias* for "swordfish" or *ziphos* for "sword"; from the Greek *cavus* for "hollow," and

from the Latin *rostrum* for "beak."

ZONES 1 TO 5, 7

DISTINCTIVE FEATURES: Forehead sloping to a relatively short, poorly defined beak; two conical teeth only, at tip of lower jaw, exposed in males, unerupted in females; not found in polar waters, but otherwise cosmopolitan.

DESCRIPTION: Maximum known length is 7.5 m. Some records of larger animals probably resulted from measurement errors or mis-identification. Females attain sexual maturity at an average length of about 6 m; males, about 5.5 m. At birth this whale is about 2.7 m long.

The body is long and robust, with a relatively small head. The angle of the forehead is not steep, sloping gradually so that the short beak is not sharply defined. The profile of the head may have a slightly concave, or scooped, appearance. With age the beak becomes less distinct. The lower jaw extends beyond the upper. The peculiar contour of the mouthline, in combination with the profile of the head, produces an appearance that has been likened to a goosebeak. The blow may project slightly forward and to the left, and it is often low and inconspicuous, even after a long dive and under conditions of good visibility. The V-shaped pair of throat grooves characteristic of beaked whales is present.

Flippers are small, with an angular outer edge. The dorsal fin can be as high as at least 38 cm, and can be smoothly falcate or low and triangular; it is located well behind the midpoint of the back. There is usually no distinct notch between the flukes, whose rear margin is somewhat concave.

Coloration can differ greatly, and should not be used as a key to identification. The back may be dark rust brown, slate gray, or fawn colored. The belly is usually lighter, although some animals are dark in both regions. The head is often paler than the rest of the body, especially (it seems) in old males, which become almost completely white on the dorsal surface anterior to the dorsal fin. The back and sides are usually covered with linear scars (attributed to intraspecific fighting), and the belly and sides with white or cream-colored oval blotches.

The single pair of conical teeth is located at the tip of the lower jaw. In females they are slender and pointed, and do not normally pierce the gum. The more massive teeth of males erupt fairly early in life, and are not concealed by the upper jaw. They are often badly worn.

NATURAL HISTORY: Though groups of as many as 25 animals have been reported, *Ziphius* is more often observed in tight schools of three to ten. Except for occasional individuals, which seem to be solitary bulls, no generalizations can be made about group composition.

Data on seasonal distribution are inconclusive, though the species is apparently a year-round inhabitant of at least some parts of its range (e.g., off New Zealand, the British Isles, western North America, and Japan).

It is rarely observed at sea, and apparently is wary of boats. Breaching has been observed, but probably happens infrequently. This is presumably a deep-diving whale, able to submerge for at least thirty minutes. The flukes are often raised as the animal begins its deep, vertical dive.

Squid and deepwater fish are the main food items. Individual strandings, sometimes of live animals, are more frequent for this whale than for any other beaked whale. These whales are believed to live for at least 35 years.

DISTRIBUTION AND CURRENT STATUS: Cuvier's beaked whale is one of the most widely distributed cetaceans, and stranding records suggest it is more common than the lack of sightings might lead one to believe. Stranded specimens have been noted from Cape Cod and the North Sea south to Tierra del Fuego and the Cape of Good Hope in the Atlantic, and from the southern Bering Sea south to Australia and New Zealand in the Pacific. It is relatively common in the Mediterranean and Caribbean Seas and the Sea of Japan. It also inhabits much of the Indian Ocean. It appears to avoid only very high latitudes in both hemispheres, though a summer sighting in the subantarctic (52° S) of New Zealand has been reported. Records from Hawaii and the Midway Islands, as well as sightings in the eastern tropical Pacific, demonstrate that it ranges far from continental land masses, and in subtropical and, at least occasionally, tropical waters.

CAN BE CONFUSED WITH: There is every likelihood of confusing this whale with bottlenose whales in either hemisphere. Perhaps the most reliable field distinction is shape of the forehead (much more bulbous in *Hyperoodon*), and length and character of the beak (longer and more well-defined in *Hyperoodon*). It can also readily be confused with nearly any species of *Mesoplodon,* and possibly with small representatives of *Berardius.* The relative shortness of the *Ziphius* beak, and the placement of its teeth (males only) at the exposed tip of the mandible, usually offer the best way to distinguish it. Remember also that the teeth of *Ziphius* are conical or oval in cross section, rather than flattened or laterally compressed, as in *Mesoplodon* and *Berardius.* Older individuals that are mostly white on the back could be confused with white whales, but the presence of a dorsal fin on Cuvier's beaked whale should rule out the finless white whale.

Tasman Beaked Whale

Tasmacetus shepherdi
Oliver, 1937
DERIVATION: *Tasma* refers to
Tasman Sea, *cetus* from the
Greek *ketos* or Latin *cetus* for
"whale"; G. Shepherd, Curator

of the Wanganui Museum in
New Zealand, obtained the
type specimen.

ZONES 3, 4, AND 5

DISTINCTIVE FEATURES: Has many standard features of beaked whales, but dentition is unique; upper and lower jaws lined with many (17-29) small, conical teeth; two much larger teeth at tip of mandibles, unerupted in females; limited to temperate waters of southern hemisphere.

DESCRIPTION: A live Tasman beaked whale has yet to be observed and positively identified. Descriptions come exclusively from a few carcasses. Lengths of measured specimens have ranged from 6 to 7 m. The body is rather robust, with contours similar to those of most other ziphiids.

The forehead is rounded and sharply set off from a long, narrow beak. The line of the mouth is straight. The throat has the usual pair of V-shaped creases. The crescentic blowhole on top of the head is asymmetric, the right margin extending farther anteriorly than the left. A moderately falcate dorsal fin is set more than two-thirds of the body length behind the tip of the beak. Flippers are narrow and short. The flukes have no median notch.

Since no living specimens have been identified and observed, nothing is known with certainty about pigmentation. In general the back is thought to be darker than the belly. There may be light lateral striping, and the head in front of and above the eyes may be whitish (as in *Ziphius* and *Hyperoodon*).

The slender beak is lined with sharp, functional teeth, rows of 17 to 21 in the upper jaw, and 17 to 29 in the lower. At the end of the mandible is a pair of much larger teeth. These apparently pierce the gum only in males.

NATURAL HISTORY: Almost nothing is known of this species' natural history. The few stomach contents that have been examined suggest that it may feed on the bottom in fairly deep water. A female 6.6 m long was sexually mature. A live animal thought to be of this species was watched briefly from shore. It surfaced twice, then disappeared; no blow was visible.

DISTRIBUTION AND CURRENT STATUS: Specimens have been reported from South Australia, New Zealand, Argentina, and Chile. The Tasman beaked whale may have a circumpolar distribution in temperate waters of the southern hemisphere. Nothing is known about its population size, and it has never been exploited by man.

CAN BE CONFUSED WITH: At sea this species might be confused with other beaked whales.

THE MESOPLODONTS

Genus *Mesoplodon.*

DERIVATION: from the Greek
mesos for "middle," *hopla* for
"arms," and *odon* for "tooth,"
meaning armed with a tooth in
the middle of the jaw.

THIS GENUS includes many obscure and taxonomically confusing
species of marine mammals, some known only from a few specimens, others only from fortuitously discovered bone fragments.
Because so little is known about most of the species and because
they appear to have many characteristics in common, they are described here as a group. The distinctions between them, and the
few features of their natural history known to us, are included in
the truncated species accounts that follow.

DESCRIPTION: All species of *Mesoplodon* have spindle-shaped bodies
that taper noticeably at both ends. The body is generally taller than
it is wide, that is, laterally compressed. Maximum length is about
6.5 m. These whales have small heads with rather well-defined
beaks. Their foreheads are not as bulbous or bluff as those of
Hyperoodon or *Berardius*. The tip of the lower jaw often extends
slightly beyond that of the upper jaw. Functional teeth are usually
present in the mandibles only, and they regularly protrude above
the gumline in adult males, but not in females. The single pair of
fully developed teeth is shaped and positioned in a distinctive manner for each species. In general, these teeth are laterally compressed
or flattened, and can thus be distinguished from the more nearly
conical teeth of *Hyperoodon* and *Ziphius,* but there are exceptions to
this rule; so it is a useful, but not infallible, generic key. The teeth of
Berardius are laterally compressed, but this genus has four rather
than two of them, and can thus be easily distinguished from all
forms of *Mesoplodon.*

The *Mesoplodon* tooth usually has a small, sharp denticle of dentine, which stands out in appearance as a "tooth upon a tooth." In adult males of some species this pointed tip is partially or completely worn away. Its position is usually near the apex of the tooth. Although dentition alone can be diagnostic for males, distinguishing females and young of the different species is very difficult. Rudimentary, functionless teeth are frequently found loosely resting in the gums of either jaw.

The blowhole usually consists of a broad half-circle, with the open side facing foward. It is not always symmetric.

The two gular, or throat, grooves characteristic of all beaked whales are present, arranged in the usual pattern of a forward-pointing, incompletely closed V.

All species have a triangular or falcate dorsal fin situated behind the center of the back. The typical *Mesoplodon* flukes have no median notch, and the rear margin is straight or slightly concave. Flippers are proportionately small and narrow, and have a somewhat pointed appearance. They are usually pressed close to the body, resting in depressions sometimes referred to as "flipper pockets."

Coloration patterns have usually been inferred from dead animals; since cetacean coloration becomes obscured after death, published descriptions should be viewed with caution. However, the linear scars or scratch marks characteristic of other beaked whales are clearly present on *Mesoplodon*, as are the oval or crescentic scars attributed to cookie-cutter sharks or lampreys. Males are

An unidentified beaked whale, probably a mesoplodont. Examination of photos from this sighting convinced several world authorities that the animal was probably M. hectori *or* M. layardi. *(Southwest of Chatham Islands, 46°43'S, 172°15'W, December 18, 1980: F. Kasamatsu. Japan Whaling Association.)*

usually more heavily scarred than females.

Some generalizations about dentition allow gross distinctions between species of *Mesoplodon*. In three species the mandibular teeth are apical, that is, placed at the tip of the lower jaw; these are *hectori, pacificus,* and *mirus.* In tooth placement these three species are similar to *Hyperoodon* and *Ziphius* (*Berardius* and *Tasmacetus* have apical positioning of their largest teeth, but both have additional developed teeth behind this anteriormost pair). Five species have the teeth set far back in the mandible, at or even posterior to the end of the mandibular symphysis; these are *bidens, densirostris, layardii, grayi,* and *ginkgodens.* Teeth in the remaining four species—*europaeus, carlhubbsi, bowdoini,* and *stejnegeri*—are positioned somewhere between the forward and rear extremes of the mandibular symphysis.

Our paintings and descriptions of many of the mesoplodonts suffer from the scarcity of photographs, specimens, and responsible accounts or drawings.

CAN BE CONFUSED WITH: At sea, virtually all species of *Mesoplodon* can be confused with one another and with *Tasmacetus* or *Ziphius.* Sometimes adult males (e.g., *M. layardii, M. stejnegeri*) can be recognized by their massive teeth that protrude outside the mouth. Also, the choices can sometimes be narrowed on the basis of locality. The much larger size of *Berardius* and the bulbous forehead of *Hyperoodon* should help distinguish them from mesoplodonts. With stranded specimens, take care to observe tooth size, shape, and positioning. Dentition often offers the only way to separate the beaked whales prior to museum preparation.

Some guide books have attempted to use suspected differences in distribution to identify beaked whales seen at particular locations and times. Although it may some day be possible to this reliably, we emphasize that existing data are sketchy, and that strandings often may tell more about the distribution of ocean currents relative to shore contours, and the distribution of interested cetologists, than about distribution of the animals in question.

Blainville's Beaked Whale

Mesoplodon densirostris
(de Blainville, 1817)
DERIVATION: from the Latin
densus for "thick, dense," and
rostrum for "beak."

ZONES 1 TO 5

DISTINCTIVE FEATURES: A massive pair of teeth set behind the mandibular symphysis, tilting forward on raised mandibles; temperate and tropical distribution, absent from polar region.

DESCRIPTION: Maximum known length in both sexes is slightly more than 4.7 m. Length at birth is probably between 1.9 and 2.6 m. There is a high arching prominence near the corner of the mouth that gives it a characteristic contour. A massive tooth is rooted at the front of this prominence on either side of the lower jaw. This characteristic is very much exaggerated in adult males, in which the teeth erupt and tilt slightly forward. The teeth of these larger males are sometimes encrusted with barnacles. The head is often flattened directly in front of the blowhole in this species, a factor that may aid in field identification. Prior to museum preparation, females and young males are difficult to distinguish from other forms of *Mesoplodon*.

    Color has been described as black or charcoal gray on the back, slightly lighter on the abdomen. Grayish white or pink blotching is typical, and scars and scratches usually cover the body. Males are

A more bizarre face would be hard to find in the animal world: an adult female Blainville's beaked whale, stranded at Cape Hatteras, North Carolina, and shown here en route to the U.S. National Museum. (James G. Mead.)

A young, relatively unscarred Blainville's beaked whale remains near one adult, while two fellow pod members glide by. (Off Hawaii: © Bill Curtsinger.)

generally more heavily scarred than females, and often have large white or reddish patches on the head. Flippers are lighter than the back; flukes, dark above and light below.

NATURAL HISTORY: Little is known, but age at sexual maturity has been estimated as nine years.

DISTRIBUTION AND CURRENT STATUS: This species may be the most widely distributed mesoplodont, having been recorded from tropical and warm temperate waters of all oceans. On the North American east coast, it has stranded from Nova Scotia south to Florida and the Bahamas, and in the Gulf of Mexico. The only European records are from Madeira and the Mediterranean Sea. There are records from the Cape of Good Hope, the Seychelles, Mauritius, and Nicobar in the Indian Ocean; northeastern Australia, Tasmania, and the northern Tasman Sea in the South Pacific; and Taiwan, Japan, and the Midway Islands in the subtropical mid-Pacific. The only record for the west coast of North America is a recent stranding in northern California. Sightings of this species have been made in Hawaii, and one stranding is known to have occurred there.

CAN BE CONFUSED WITH: Only adult male Blainville's beaked whales are likely to be distinguishable at sea from other mesoplodonts. The high arching contour of the mouthline, with a massive forward-tilting tooth on either side, distinguishes them from all other mesoplodonts except Stejneger's beaked whale and perhaps the ginkgo-toothed beaked whale. As with all mesoplodonts, one ought to be conservative in field identification of Blainville's beaked whale, though its confirmed wide distribution makes it a suspect in virtually any sighting of a *Mesoplodon* at sea.

Sowerby's Beaked Whale

Mesoplodon bidens
(Sowerby, 1804)
DERIVATION: from the Latin *bis*
for "two," and *dens* for "tooth."

ZONES 2 AND 8

DISTINCTIVE FEATURES: Mandibular teeth appear near the middle of beak; known only from temperate to subarctic North Atlantic.

DESCRIPTION: This whale reaches lengths of at least 5 m. Length at birth is about 2.4 m. A 2.7 m calf weighed 185 kg. In some photographs the head has appeared to have a prominent bulge in front of the blowhole, a slightly concave forehead, and a moderately long beak, but in others has not appeared to differ appreciably from that of other mesoplodonts. The flippers (one-seventh to one-ninth of body length) are considered long for a *Mesoplodon*, but still small by general cetacean standards.

Descriptions of coloration, based largely on dead specimens, are not consistent. Adults are generally a dark charcoal gray, perhaps on the belly as well as the back, although some have been said to be much lighter ventrally. Light spots, sometimes described as "splashes" or "smears," are distributed irregularly over the body. Younger specimens seem to have noticeably lighter bellies and fewer spots.

The two mandibular teeth are situated at the middle of the

snout, near the posterior end of the mandibular symphysis. In adult males they project backward, then slightly forward, and are visible outside the mouth.

NATURAL HISTORY: Individuals and pairs have been known to strand alive. Some very limited data suggest that mating and birth usually occur in late winter and spring.

DISTRIBUTION AND CURRENT STATUS: This is the most northern species of *Mesoplodon* in the Atlantic. Its range covers only the cold temperate to subarctic North Atlantic, and it appears to be much more common in European than American waters. Strandings in Massachusetts and Newfoundland are the only evidence of its presence in the western North Atlantic. In addition, there is a specimen in the United States National Museum from Cartwright, Labrador. On the other side, it has been reported from Iceland, Norway (north of Trondheim Fjord), Sweden, the Baltic Sea coast of Germany, Holland, Belgium, France, the British Isles, and as far south as Madeira. Stranding records indicate that Sowerby's beaked whale is relatively common in the North Sea. Contrary to some older references, this species has never been reliably reported in the Mediterranean Sea.

CAN BE CONFUSED WITH: The fact that this animal's range is limited to the North Atlantic should obviate confusion with most other species in the genus. However, it cannot reliably be distinguished from other North Atlantic mesoplodonts – *M. densirostris, M. europaeus, M. mirus,* and possibly *M. grayi* – unless a specimen is examined by a competent zoologist. Adult male Blainville's beaked

Since there is no major fishery for mesoplodonts, most knowledge about them comes from strandings. This is a female Sowerby's beaked whale. (On the English Channel at Colleville/Mer, Calvados, France, September 1975: Daniel Robineau.)

whales (*M. densirostris*) have massive teeth and a distinctive arched contour of the mouthline, and the teeth of True's beaked whale (*M. mirus*) are at the tip of the lower jaw; so the adult males of these two species can be distinguished from Sowerby's beaked whale by means of external characteristics.

The range of Sowerby's beaked whale overlaps that of the northern bottlenose whale. Young bottlenose whales in particular might be confused with adult Sowerby's beaked whales. The adult bottlenose whale, however, is nearly twice as large as the adult Sowerby's beaked whale, and the former's bulbous melon is a distinguishing feature.

Gervais' Beaked Whale

Mesoplodon europaeus
Gervais, 1855
DERIVATION: *europaeus* refers to
Europe, where the type
specimen was found floating in
the English Channel during the
1840s.

ZONES 2 AND 5

DISTINCTIVE FEATURES: Mandibular teeth about one-third of the way from tip of snout to corners of mouth; known only from the Atlantic in warm temperate to tropical waters.

DESCRIPTION: Maximum length is probably about 5 m; physically mature adults usually are 4.5 to 4.8 m long. Length at birth is probably close to 2.1 m. The head of this species is proportionately small, the beak narrow. Color is basically dark gray on the back and sides, lighter ventrally, with irregular white markings on the ventrum, especially in the genital region.

The two mandibular teeth erupt in adult males about one-third of the way back from the tip of the snout to the corners of the mouth. They protrude outside the mouth of adult males, and fit into grooves in the skin of the outer upper jaw.

NATURAL HISTORY: Little is known, but minimum longevity has been estimated as 27 years.

DISTRIBUTION AND CURRENT STATUS: This species is known only

One of two female Gervais' beaked whales brought ashore by local fishermen. (Bulls Bay, Jamaica, February 1953: J. J. Rankin.)

---

from the Atlantic, where it appears to favor warm temperate and subtropical waters. It may be closely associated with the Gulf Stream. The type specimen was found floating in the English Channel. Since then, strandings have been documented at Guinea-Bissau, West Africa, and on Ascension Island, south of the equator. All other records come from the east coast of North America, from Long Island (New York) south to Florida, including the Gulf of Mexico as far west as Texas, or from the Caribbean region, including Jamaica, Cuba, and Trinidad.

CAN BE CONFUSED WITH: The range of Gervais' beaked whale regularly overlaps with that of True's beaked whale. Tooth position, at the tip of the mandibles in True's beaked whale, is the best clue for distinguishing the two. Gervais' beaked whale probably has a more southern, warm-water distribution, which may also help in identification. Blainville's beaked whale may pose problems as well, since its teeth are positioned similarly to those of Gervais' beaked whale. However, the former has an arched contour of the mouthline and more massive teeth, which should make at least adult males distinguishable from Gervais' beaked whales.

Cuvier's beaked whale is found in the same latitude as Gervais' beaked whale, and may be a source of confusion. The former has conical teeth situated at the tip of the lower jaw; so dentition can aid in identification. Also, adult Cuvier's beaked whales probably have paler pigmentation, especially on the head, than Gervais' beaked whales.

True's Beaked Whale

Mesoplodon mirus
True, 1913
DERIVATION: from the Latin
mirus for "wonderful."

DISTINCTIVE FEATURES: Two teeth at tip of mandibles, exposed outside mouth in adult males; known only from the temperate Atlantic and southwestern Indian Oceans.

DESCRIPTION: Maximum known length is between 5 and 5.5 m. The body is shaped much like that of Cuvier's beaked whale, being chunky in the midriff and narrowing rapidly toward the tail. There is a slight indentation in the area of the blowhole, a slight bulge to the forehead, and a pronounced beak.

Color is dull black to dark gray on the back, lighter slate gray on the sides, and gray on the belly. Light spots or blotches are usually present, especially in the genital and anal regions, as are the scratches apparently made by the teeth of males.

The two mandibular teeth rest at the tip of the snout, which makes this species especially difficult to distinguish from Cuvier's beaked whale. They angle forward and, in males, can be seen outside the mouth.

DISTRIBUTION AND CURRENT STATUS: This species is known only

from strandings in the temperate North Atlantic and in southeastern Africa. Until very recently it was believed to dwell only in the North Atlantic, but in 1959 a specimen was found in Cape Province, South Africa. Several more South African records have been reported since then. The individuals present off the southern coast of Africa may belong to a geographically separate stock, or the species' range may be much wider than has been estimated from confirmed stranding records. Strandings have occurred between northeastern Florida and Nova Scotia in the western North Atlantic, and in the British Isles in the eastern North Atlantic.

CAN BE CONFUSED WITH: This species is often confused with Cuvier's beaked whale because of tooth placement and overall appearance, less often with bottlenose whales (*Hyperoodon* spp.) for the same reasons. Its range overlaps significantly those of *M. bidens, M. europaeus,* and *M. densirostris,* but in all these the teeth are well behind the tip of the snout.

Strap-toothed Whale

Mesoplodon layardii
(Gray, 1865)
DERIVATION: E. L. Layard was
curator of the South African
Museum, 1855–1872; he pro-
vided the drawings on which
Gray based his description of
the species.   ZONES 3 TO 5

DISTINCTIVE FEATURES: Unique "wraparound" dentition of adult
males; known only from temperate southern hemisphere.

DESCRIPTION: Maximum length is slightly more than 6 m. The
color has been described as dark purplish brown or dark gray to
black above, shading to white (sometimes with yellowish tones)
below. Fresh specimens described by South African scientists Peter
Best and Graham Ross seem to show that this species has a complex
color pattern, marked by white zones around the genital slit, below
and behind the flippers, and around the face, which is dark. There
is also a grayish lightening of the dorsal surface in front of the dorsal
fin. The back and flanks often have large, irregular white patches.
    The most interesting and diagnostic feature of this species is the
strap-shaped pair of teeth that emerges from the mandibles, well
behind the tip of the beak, in males. The teeth extend upward and
backward beside the upper jaw, eventually curling over it and preventing the mouth from opening fully. Maladaptive as such structures would appear to be, they apparently do not interfere with the
eating process, and may even act as guide rails to hasten prey toward

the throat. In females and probably young males the teeth are not apparent, making identification at sea difficult.

NATURAL HISTORY: Most strandings in New Zealand have occurred in summer, but there is no evidence of seasonality in the species' presence around Australia and Tasmania.

DISTRIBUTION AND CURRENT STATUS: The species appears to be circumpolar in the southern hemisphere; there are stranding records from South Africa, Uruguay, the Falkland Islands, Tierra del Fuego, New Zealand, southeastern Australia, and Tasmania. Its northern limit is believed to be about 30° S, and there is no reason to think its range extends to Antarctic waters.

CAN BE CONFUSED WITH: Thanks to their bizarre dentition, adult males of this species are among the few types of *Mesoplodon* that can be distinguished readily from all other whales, sometimes even at sea. Females and young males, however, will probably be very difficult to distinguish from other species of beaked whales.

Gray's Beaked Whale

Mesoplodon grayi
von Haast, 1876
DERIVATION: J. E. Gray, who
died in 1876, had been Director
of the British Museum.

ZONES 2 TO 5

DISTINCTIVE FEATURES: Teeth relatively small and triangular, placed near middle of long, narrow snout; limited to temperate latitudes of southern hemisphere and the eastern North Atlantic.

DESCRIPTION: Maximum length is probably 5.5 to 6 m, and calves are less than 2.4 m long at birth.

The head is small, the beak long and narrow. Color has been described inconsistently by different authors: the dorsal surface may be dark brownish gray to black; the sides gray or a mottled brownish gray; the belly light gray or white. The beak and throat are often white or white-flecked, and there may be conspicuous white markings around the navel, genital aperture, and anus.

The two mandibular teeth are located well behind (about 20 to 24 cm) the tip of the snout. They are triangular in shape and less massive than in most other beaked whales, and erupt only in adult males. Their attitude is upright, with no noticeable inclination either forward or backward. In addition to these two mandibular teeth, Gray's beaked whale normally has rows of 17–22 small maxillary teeth on each side of the upper jaw. They are conspicuous

At least 53 Gray's beaked whales, including this female, are known to have stranded on New Zealand beaches between 1873 and 1973. (Waverly Beach, New Zealand, September 25, 1931: courtesy of Alan N. Baker.)

enough that one early observer proposed a new genus, *Oulodon*, for this species, because of the consistent presence of functional maxillary teeth.

NATURAL HISTORY: Dale W. Rice, who observed and identified this species on several occasions in the Indian Ocean, referred to its "peculiar habit of sticking its long needle-like white snout out of the water as it breaks the surface to blow."

Stranding is fairly frequent in New Zealand. A mass stranding of 28 animals was reported at the Chatham Islands in the nineteenth century. Such an occurrence is unusual for beaked whales.

DISTRIBUTION AND CURRENT STATUS: This species appears to be circumpolar in the southern hemisphere. Most stranding records are from New Zealand; so it may be especially common there. Strandings have also occurred at the Chatham Islands and in South Australia, Argentina, and South Africa. Sighting frequency suggests that the species may be common in the Indian Ocean south and east of Madagascar in waters deeper than 1,000 fathoms. A well-documented record of an animal that stranded on the North Sea coast of the Netherlands in 1927 constitutes the only evidence for its presence in the northern hemisphere.

CAN BE CONFUSED WITH: It is probably difficult to distinguish this species from other sympatric species of *Mesoplodon* at sea, although several experienced cetologists feel they can recognize adult males. The mouthline is straight compared to those of Blainville's beaked whale and the ginkgo-toothed beaked whale. Young male strap-toothed beaked whales in early stages of tooth eruption and development may look similar to adult Gray's beaked whales. Stranded adult males might be identifiable by tooth size and placement, but museum preparation is necessary to identify females and subadult males reliably.

Andrews' Beaked Whale

Mesoplodon bowdoini
Andrews, 1908
DERIVATION: George S.
Bowdoin was a trustee of the
American Museum of Natural
History who helped enlarge the
museum's cetacean collection.   ZONES 3 AND 4

DISTINCTIVE FEATURES: A pair of massive teeth in males, located slightly behind the mandibular symphysis; known only from temperate Indian and Pacific Oceans.

DESCRIPTION: Maximum known length is close to 4.6 m. Very little is known about the external appearance of this species, but its skull and skeleton closely resemble those of *M. carlhubbsi*. The two species may prove to be identical, with only subspecific differences between the northern- and southern-hemisphere populations.

The two mandibular teeth are situated approximately at mid-beak, just behind the mandibular symphysis, set in slightly raised sockets. They probably protrude outside the mouth in adult males, and remain concealed in the gum in females and young. Although *M. grayi* has similar tooth positioning, its teeth are much less massive and conspicuous than those of *M. bowdoini*.

DISTRIBUTION AND CURRENT STATUS: Evidence of stranding by this species comes from New Zealand (including Campbell Island), Tasmania, Western Australia, Victoria (eastern Australia), and Kerguelen Island; so it appears to be confined to temperate waters of the South Pacific and Indian Oceans.

CAN BE CONFUSED WITH: Andrews' beaked whale is very difficult to distinguish from Hubbs' beaked whale and Stejneger's beaked whale of the North Pacific. Identification should be left to specialists familiar with the slight differences between these species. To date, the range of *M. bowdoini* has not been found to overlap those of *M. carlhubbsi* and *M. stejnegeri*. Although Gray's beaked whale (*M. grayi*) has tooth positioning similar to that of Andrews' beaked whale, its teeth are less massive.

Longman's Beaked Whale*

Mesoplodon pacificus
Longman, 1926
DERIVATION: Unclear; Longman probably named it *pacificus*
to distinguish it from the similar
M. mirus, "which is only known from three specimens
obtained from the Atlantic."

ZONE 3

DISTINCTIVE FEATURES: Skull similar to that of *M. mirus;* apparently found only in Indian and South Pacific Oceans.

DESCRIPTION: This may be the least-known cetacean, since evidence for its existence rests on only two weather-worn skulls found
in widely separate localities. Its taxonomic status is not fully resolved. Various systematists have considered it to be: (1) a misidentification of *Hyperoodon planifrons,* (2) a subspecies of *M. mirus,* or (3)
in a separate genus called *Indopacetus.*

DISTRIBUTION AND CURRENT STATUS: One specimen came from
Queensland, Australia, the other from Danane, Somalia. The
species may be limited to the Indo-Pacific.

---

* Since no photographs exist of Longman's Beaked Whale, a responsible
illustration could not be produced.

Hector's Beaked Whale

Mesoplodon hectori
(Gray, 1871)
DERIVATION: J. Hector was curator of the Colonial Museum in Wellington, New Zealand, where the type specimen originated.

ZONES 3 TO 5

DISTINCTIVE FEATURES: Mandibular teeth near tip of lower jaw; apparently limited to southern hemisphere and North Pacific.

DESCRIPTION: This species is known from only a few specimens. A 3.9-m male and a 4.43-m female were found to be physically mature, but both sexes probably reach lengths of 4.5 m. Color is basically dark above and light below. The back may have a grayish brown cast, and the chin and lower jaw may be pale gray. The flanks can have linear and oval scars, as in most beaked whales. Flukes of adult males are white on the undersides, and there can be a white area in the vicinity of the umbilicus.

The two mandibular teeth are relatively small, triangular, and flat; they are placed near the tip of the lower jaw. One zoologist proposed that *M. hectori* was a spurious species, and that the skulls were in fact young specimens of *Berardius arnuxii;* but this view has been repudiated, and *M. hectori* accepted as a valid species.

DISTRIBUTION AND CURRENT STATUS: Existing records suggest that this whale is confined to temperate latitudes of the southern hemisphere, where it may be circumpolar, and to the temperate eastern North Pacific. Specimens have come from Tierra del Fuego, New Zealand, Tasmania, South Africa, and the Falkland Islands in the southern hemisphere, and from southern California in the North Pacific.

CAN BE CONFUSED WITH: Because of its tooth positioning, *M. hectori* easily may be confused in the southern hemisphere with *M. mirus, M. pacificus*, or young of *Berardius arnuxii*. In the North Pacific, it is the only mesoplodont with teeth far forward in the lower jaw, well ahead of the mandibular symphysis.

Especially because of its relatively short beak and the position of its teeth, Hector's beaked whale may often be confused with Cuvier's beaked whale. The latter's larger body size and cone-shaped rather than flattened teeth are likely to help distinguish the two species.

Ginkgo-toothed Beaked Whale

Mesoplodon ginkgodens
Nishiwaki and Kamiya, 1958
DERIVATION: Ginkgo refers to the ginkgo tree *(Ginkgo biloba)*, whose leaves are shaped like the teeth of this species; from the Latin *dens* for "tooth."

ZONES 1 AND 3

DISTINCTIVE FEATURES: Mandibular teeth close to midbeak; limited to Indo-Pacific.

DESCRIPTION: Maximum length is probably close to 5 m; weight, 1,500 kg. Pigmentation cannot be described with confidence, since living specimens have not been examined (the general body color of dead specimens was described as midnight black, with the belly being somewhat lighter). There are many oval white scars on the belly and sides.

There is a raised area in the rear half of each side of the lower jaw. The mandibular teeth, which are shaped like the leaf of the ginkgo tree, rest in the front half of these plateaus. The teeth of this species are the widest (100 mm or more) anteroposteriorly of any known *Mesoplodon*.

DISTRIBUTION AND CURRENT STATUS: The ginkgo-toothed beaked whale is known only from the warm temperate and tropical North Pacific and the northern Indian Ocean. Existing records are from southeastern Japan, Taiwan, Sri Lanka, and southern California.

A fisherman found this male ginkgo-toothed beaked whale on the beach and sold it at the Ito fish market. The small white circle on the raised mouthline is an erupted tooth. (Near Ito, Shizuoka Prefecture, on Sagami Bay, Japan, July 28, 1971: Toshio Kasuya.)

There is only one confirmed record from the coast of North America (at Del Mar, California), and the species is believed to be more common in the western North Pacific than elsewhere. Some hunting of this species apparently takes place in Taiwan.

CAN BE CONFUSED WITH: In the North Pacific and Indian Oceans, where it is known to be present, the ginkgo-toothed beaked whale can be confused with other species of *Mesoplodon* (mainly *M. densirostris*, *M. carlhubbsi*, and *M. stejnegeri*) and with Cuvier's beaked whale. In males the exposed mandibular teeth and contour of the mouthline can help rule out *Ziphius*. Male Hubbs' beaked whales (*M. carlhubbsi*) have the distinctive white "beanie" on the head; so they can be distinguished. Positive identification of females and immatures requires careful examination by a specialist.

Stejneger's Beaked Whale

Mesoplodon stejnegeri
True, 1885
DERIVATION: Leonhard H.
Stejneger, formerly curator of
the U. S. National Museum,
collected the type specimen in
1883.   ZONES 1 AND 7

DISTINCTIVE FEATURES: A single pair of teeth, erupted well above the gum in adult males, set far behind tip of snout; males have high prominences at corners of mouth; limited to northern North Pacific.

DESCRIPTION: Adults grow to at least 5.3 m and probably 1,500 kg. Adult males have a very conspicuous elevation on the lower jaws, which begins well behind the tip of the snout. A massive tooth protrudes through the gum on the forward edge of this rise, very near the peak. The exposed portion of the tooth is tilted forward slightly, with the denticle placed at the anterior upper edge of the tooth. In dorsal view these teeth appear to converge, pinching either side of the rostrum. They are often noticeably worn along the forward edge. The mouthline of females (and presumably young males) is much straighter, although it has a somewhat sinusoidal curve to it. Stejneger's beaked whale has no other appreciable external differences from the other species of *Mesoplodon*.

Very few of these whales have been seen alive; so little can be said about their color pattern. However, some individuals are

The armament of an adult male Stejneger's beaked whale. With its mouth shut tightly, this animal can be formidable; it can inflict slashing wounds on companions or opponents. (Homer, Alaska, November 14, 1977: Francis H. Fay.)

known to be grayish brown on the back and lighter on the belly, with very conspicuous light, off-white areas on the sides behind the head, at the neck, and around the mouth. The post-mortem darkening of stranded whales belies the handsome marking of this species in life. Adults often have many oval white scars on the flanks and in the genital region, and males in particular have long scratch marks on much of the body.

DISTRIBUTION AND CURRENT STATUS: Stranding records indicate that Stejneger's beaked whale is an inhabitant of the cold temperate to subarctic North Pacific. It apparently is distributed across much of the southern Bering Sea, south to the northern Sea of Japan in the west and Monterey, California, in the east. Strandings are fairly common on the Aleutians; so the species may be especially common in that area.

CAN BE CONFUSED WITH: Stejneger's beaked whale is most likely to be confused with Hubbs' beaked whale, although their ranges appear to overlap only between British Columbia and California. As far as is known, examination by a specialist, perhaps after museum preparation of the skull, is needed to identify females and juveniles. The very prominent white "cap" on the head of adult male Hubbs' beaked whales, however, appears to be a diagnostic feature, and can help distinguish them from Stejneger's beaked whales.

Cuvier's beaked whales share the range of Stejneger's beaked whales; the two are likely to be confused at sea, and perhaps even on the beach. The former has a shorter, less well-defined beak, and grows much larger. Adult males can be distinguished by dentition, since the two exposed teeth in Cuvier's beaked whale emerge at the tip of the lower jaw and are conical in shape.

Hubbs' Beaked Whale

Mesoplodon carlhubbsi
Moore, 1963
DERIVATION: Carl L. Hubbs
(1894–1979) was an eminent
American marine zoologist.

ZONE I

DISTINCTIVE FEATURES: Adult males have a distinctive white prominence, or "beanie," anterior to the blowhole, and raised areas in the rear half of each lower jaw that house large straplike teeth, and are white on the front half of the beak; they are limited to the North Pacific.

DESCRIPTION: Maximum known size is about 5.3 m and 1,500 kg. Newborn are probably about 2.5 m long. The skull and skeleton closely resemble those of *M. bowdoini*. By far the most distinctive feature, found only in adult males, is the stark white convexity in front of the blowhole. It is reminiscent of a cap or beanie. The mouthline of adult males has prominences on either side that support the single pair of flattened teeth. These teeth are fairly massive, and are exposed outside the closed mouth. Viewed dorsally, they appear to pinch the rostrum. Females and young have a much less dramatically curved mouthline, and their teeth are unerupted.

Hubbs' beaked whale differs little in other ways from other species of *Mesoplodon*. The proportionately small flippers fit into slight depressions (flipper pockets) when pressed against the body.

Little of the body of this male Hubbs' beaked whale has been left unscarred. (Drakes Bay, Marin County, California, March 19, 1950: Woody Williams.)

The falcate dorsal fin, situated behind the midback, is about 22 to 23 cm high in most adults.

General body color is dark gray to black; females may be somewhat lighter laterally and ventrally. Males in particular are heavily scarred, with long (to 2 m) linear scratches and oval spots. In addition to the white "beanie," the front half of the beak is white in adult males. The rostrum and front half of the lower jaw in females and immature animals is also relatively pale.

NATURAL HISTORY: Next to nothing is known about the natural history of this species. The calving season has been estimated to be in midsummer. It probably does not congregate in large herds; groups of two to ten are likely to be usual. Stomach contents of stranded specimens suggest that they prey mainly on squid and mesopelagic fish.

DISTRIBUTION AND CURRENT STATUS: Hubbs' beaked whale inhabits the cold temperate North Pacific. Knowledge of its distribution is based solely on stranding records. Off Japan it appears to be restricted to waters off northwest Honshu. Its southern limit in the western North Pacific may be the warm Kuroshio Current, its northern limit the cold Oyashio Current, with most activity centered at about 38° N, where these two currents meet. On the eastern side the range seems somewhat broader, extending from San Diego (33° N) in the south to Vancouver Island (51° N) in the north, and related to the confluence of the Subarctic and California Current systems. Again, stranding evidence may tell more about the distribution of cetologists than of cetaceans.

CAN BE CONFUSED WITH: *M. carlhubbsi* is one of the few species in the genus for which external characteristics permit identification at sea, but only for adult males. The white "beanie" is apparently unique. Because of their overlap in range and similar dentition, *M. stejnegeri, M. ginkgodens,* and *M. densirostris* are likely to be confused. Adult males of all three species have raised prominences in the lower jaw and exposed teeth. Generally speaking, the amount of enamel exposed above the gum (and therefore outside the mouth) is greater in *M. carlhubbsi* and *M. stejnegeri.* Another useful feature is that the denticle is placed just behind the tooth's anterior edge in *M. carlhubbsi,* in a line with the tooth's front edge in *M. stejnegeri,* and centrally in *M. ginkgodens* and *M. densirostris.*

Cuvier's beaked whale may be problematic, because adult males have a white head. The beak, however, is generally less prominent and defined, the conical teeth are positioned at the apex of the mandibles, and body size is larger.

7. The Oceanic Dolphins

Family: Delphinidae

Irrawaddy Dolphin

Orcaella brevirostris
(Gray, 1866)
DERIVATION: *orcaella* is
diminutive of the Latin *orca* for
"a kind of whale"; from the
Latin *brevis* for "short," and
rostrum for "beak."

ZONE 3

DISTINCTIVE FEATURES: High, bluff forehead; no beak; broad spatulate flippers; uniform slate-blue coloration; dorsal fin low and slightly falcate; found only in the tropical Indo-Pacific, primarily near shore.

DESCRIPTION: Adults are 2 to 2.5 m long. A full-term fetus was 85 cm long. The forehead is bluff; the mouthline is fairly straight, but angles upward toward the eye; and there is a neck crease. The crescentic blowhole is located to the left of the top of the head, with the open side facing anteriorly.

Flippers are broad and relatively long, with a blunt or rounded tip. The dorsal fin is small, placed slightly behind the middle of the back, low, and with a barely concave rear margin. The tail stock is very narrow, with very little keeling; the flukes are notched and with a shallow concave trailing edge. Color is generally gray to dark slate blue, with a noticeably lighter belly.

There are normally about twelve to nineteen teeth on either side of the upper jaw, and twelve to fifteen on either side of the lower jaw. Usually there are more teeth in the upper than in the lower jaw.

Edward Mitchell has said of this dolphin, "In size, appearance, mobility of body movement and habit, it is the equatorial equivalent of the Arctic white whale . . . The anatomical similarities indicate a phyletic relationship, suggesting that this species is a relict of an antitropical species pair."

NATURAL HISTORY: This is not a noticeably gregarious species, although solitary individuals are rarely encountered. Groups usually consist of no more than ten animals. They are sometimes seen in the vicinity of *Sousa chinensis*. Life-history details are not available.

These dolphins keep close to the coast, and often ascend rivers. They apparently can live permanently in fresh water. In the Mekong River their movements are influenced by the seasonal change in water levels: the flooding that begins in June allows them to penetrate lakes and tributaries that are otherwise inaccessible.

Since they are small and quiet, and often inhabit turbid estuarine waters, some vigilance is required to locate and observe them. Often the only way to detect them is by listening carefully for the sound of exhalation. They tend to show much of the body when surfacing, which facilitates easy recognition. Maximum diving time is about ten minutes, but so long a dive is undertaken only under duress. Normally, they surface at intervals of no longer than a minute. They have been seen to make low, horizontal leaps out of the water, but such behavior apparently is unusual. This dolphin is said to spit quantities of water into the air while spy hopping, i.e., raising the head above the water as if supported on its flippers. Maximum swimming speed for short distances is said to be about 20 to 25 km per hour. Several have been kept successfully in captivity in Australia and Indonesia. Fish are known prey; crustaceans may also be eaten. Nothing is known about predation or longevity.

DISTRIBUTION AND CURRENT STATUS: This species is an inhabitant of warm, tropical waters throughout much of the Indo-Pacific. It seems to prefer coastal areas, particularly the muddy, brackish waters at river mouths, and it even ascends far up large rivers, such as the Irrawaddy in Burma, the Mahakam in Borneo, the Mekong (as far as Cambodia), and the Ganges and Brahmaputra in India.

The Irrawaddy dolphin is not as provincial as its common name implies; it is widely distributed in coastal Indo-Pacific waters. (Cairns, Queensland, Australia: William H. Dawbin.)

Records from northern Australia and New Guinea represent the apparent eastern limits of its range, which extends westward and northward throughout Indonesia and Indo-China, including most of the Indo-Malay Archipelago. The Bay of Bengal is probably the western extreme of its range. No offshore records are known; so the species is not likely to be seen beyond a few kilometers from shore.

There is not and probably never has been a major fishery for this dolphin, although there probably has been (and will be) casual local hunting by aborigines in some parts of its range. Some fishermen along the Irrawaddy and Mekong Rivers believe that *Orcaella* herds fish into their nets; so they have traditionally been inclined to protect it. Also, superstitious reverence for the species exists in Cambodia, where it apparently is fully protected. Some accidental entrapment occurs in antishark nets off Queensland, Australia, and in aboriginal fish traps elsewhere.

CAN BE CONFUSED WITH: The Irrawaddy dolphin is likely to be confused with the finless porpoise throughout much of its range. The two species are externally almost identical except that the finless porpoise is smaller, and lacks a dorsal fin, but the Irrawaddy dolphin has one.

The Irrawaddy dolphin might be confused with the Indo-Pacific hump-backed dolphin, especially east and south of Indonesia, where the latter lacks the distinctive humped dorsal fin. The larger size, relatively long beak, and more prominent dorsal fin of the hump-backed dolphin should usually distinguish it adequately.

Melon-headed Whale

Peponocephala electra
(Gray, 1846)
DERIVATION: from the Greek
pepon for "melon, gourd," and
kephale for "head"; *Electra* was a
nymph of Greek mythology.

ZONES 1 TO 5

DISTINCTIVE FEATURES: Beak absent or very subtle; anterior to blowhole, profile of top of head smoothly curved downward; bottom flatter; mouthline short and angled upward; head, from above and below, triangular; flippers long, slim, generally pointed on tips; lips often white; 21-25 teeth in each jaw; tropical.

DESCRIPTION: Melon-headed whales reach at least 2.6 m (females) to 2.7 m (males). One male was reportedly sexually mature at about 2.26 m, another at 2.27 m. Size at birth is unknown, though fetuses that appeared to be full-term in photographs were estimated to be 75 to 90 cm, and one obviously young calf was 1.2 m.

In general the profile of the head is like that of the false killer whale, but with a sharper appearance to the tip of the snout. Also, the mouthline is shorter and tilted slightly upward. The forehead is rounded, and ends in a rounded melon shape or a very indistinct beak. Often the beak is so subtle that there is only a slight indentation above the upper lip. The lower portion of the "face" is slightly concave in most animals. In profile the line of the throat is flat. From above and below the head is distinctly triangular.

A few melon-headed whales have been brought into captivity for brief periods in Japan, the Philippines (for the Hong Kong Aquarium), and Hawaii. (Sea Life Park, Hawaii, 1980: Robert L. Pitman.)

The body is elongated and usually slim, with a rather slender tail stock, similar to those of the larger false killer whale and similar-size pygmy killer whale. The flippers are long (may be more than 50 cm), slim, and generally pointed on the tip. The dorsal fin is tall (up to about 30 cm) and distinctly back-curved.

Coloration is generally uniformly black on the back and sides, and slightly lighter on the belly. There is often a subtle "cape" pattern along the dorsal midline that dips onto the side as a down-pointing triangle just below the dorsal fin. In the field one also might detect a triangular darkening in the face that tapers toward the eyes. This mask is apparently highlighted by shadowing that results from the slightly concave face and the hint of a beak on some animals. The areas around the anus, genitals, and lips are sometimes unpigmented, appearing light gray, white, or pink. There is also a light gray hint of an anchor-shaped throat patch. There are 20-25 small, sharply pointed teeth in the upper jaw, 22-24 in the lower.

NATURAL HISTORY: Melon-headed whales generally form large herds, of 150-1,500 animals, not infrequently associated with Fraser's dolphins, and sometimes with spinner dolphins or spotted dolphins.

Little is known of their reproduction. In the southern hemi-

Melon-headed whales, normally pelagic, are shown here in unaccustomed surroundings, impounded by nets and awaiting slaughter. (Taiji, Japan, February 1980: Howard Hall, Living Ocean Foundation.)

sphere, newborn have been reported in July and August, and term fetuses from an August stranding. The large number of pregnant females and newborn young found in the August stranding support the contention that there is a spring breeding season. No migration is known.

Melon-headed whales are very fast swimmers, and, when fleeing a ship, frequently bunch tightly together, working the sea surface into a froth.

Surfacing individuals often break the water at a shallow angle, creating much spray, which obscures the animal's pigmentation pattern. They sometimes jump clear of the water.

There are some curious recent reports of melon-headed whales (and possibly pygmy killer whales) herding and perhaps attacking small porpoises of the genus *Stenella* that were escaping from seine nets of Pacific tuna fishermen. The species feeds on squids and a variety of small fish.

DISTRIBUTION AND CURRENT STATUS: The species appears to be distributed worldwide in tropical and subtropical waters. Specimens and sightings have been reported from the west coast of Central America, Hawaii, the Marquesas and Tuamotus, the Philippine Sea (particularly near Cebu Island), Australia, Japan, the Maldive Islands, the Seychelle Islands, the Lesser Antilles, the tropical Atlantic, the Gulf of Guinea, and pelagic waters of the equatorial belt in the tropical eastern and central Pacific. The species is reported as abundant only in the Philippine Sea, especially near Cebu Island. Everywhere else it appears to be rare. It is pelagic, ranging from the continental shelf seaward and around oceanic islands, and appears to stay mainly in equatorial water masses. Small numbers have been taken intentionally off Japan, and incidentally in eastern Pacific seine fishing for tuna.

CAN BE CONFUSED WITH: At sea melon-headed whales might be confused with false killer whales or pygmy killer whales. They are much smaller than false killer whales, have a slightly more pointed snout, and lack the prominent hump on the leading margin of the flippers.

They are approximately the same size as pygmy killer whales, but usually lack the distinctive brownish cape, which is clearly demarcated from the slightly lighter color of the sides by a triangular zone extending below the dorsal fin, and may lack the extension onto the sides of the white area around the genitals. This species has pointed flippers, but those of pygmy killer whales are rounded on the tips. Melon-headed whales also have a slightly more pointed snout.

On the beach they can be distinguished from other species of slim blackfish by their larger number of teeth, usually 21–25 per row, whereas other blackfish have less than fifteen per row.

Pygmy Killer Whale

Feresa attenuata
Gray, 1874
DERIVATION: *Feres* is a vernacu-
lar French name for a dolphin;
from the Latin *attenuatus* for
"drawn out, tapered, thin."

ZONES 1, 2, AND 4;
PROBABLY 5 AND 3

DISTINCTIVE FEATURES: No beak; head rounded; lips white; lower
jaw and chin may be white; dark grayish brown cape on back, extending farther down side below dorsal fin; tropical and subtropical.

DESCRIPTION: Maximum reported length is 2.7 m. Males and
females are sexually mature by at least 2.2 m. Newborn are about 80
cm. The head is smoothly rounded in profile and in dorsal view,
with a slightly underslung jaw and no beak. The body is usually
slender. The flippers are slightly rounded on the tips, an important
feature for distinguishing the pygmy killer whale from otherwise
similar blackfish. The prominent falcate dorsal fin, located near the
center of the back, is usually 20 to 30 cm tall (though it may reach 38
cm in some individuals) and resembles that of the bottlenose
dolphin.

The color has been described as dark gray, brownish gray, or
black on the back. The sides are lighter gray. There is a small zone of
white or light gray on the underside, which is often wide on the
chin, between the flippers, around the genitals, and on the ventral

surface of the tail stock. The cape is indistinct and similar to that of the spinner dolphin, with its maximum width below the dorsal fin. The white zone on the chin, described as a "goatee," is often clearly visible in swimming animals. Often the lips are also white.

Pygmy killer whales have eight to eleven teeth in each side of the upper jaw and eleven to thirteen in each side of the lower jaw. Many specimens reportedly have one fewer on the right than on the left side. The teeth are smaller than those of the false killer whale, and far fewer than those of the melon-headed whale.

A herd of pygmy killer whales form a fast-moving chorus line. (Eastern tropical Pacific, 11°38′N, 94°58′W, February 8, 1980: M. Scott Sinclair.)

NATURAL HISTORY: Little is known about the species, though sightings of live animals in the eastern tropical Pacific are rapidly increasing, and several individuals have been maintained in captivity. Pygmy killer whales may form herds of several hundred individuals, though groups of fifty or fewer are far more common. They have been found with Fraser's dolphins. They are quick and lively, often wary of boats, and tend to bunch together when fleeing the cause of a disturbance.

In captivity they are aggressive, and elicit fright reactions from captives of other species. They have been reported to attack other small cetaceans in the South Atlantic and tropical Pacific.

DISTRIBUTION AND CURRENT STATUS: The pygmy killer whale is apparently widely distributed in tropical waters. It has been encountered off Florida and the West Indies, and in the Mediterranean Sea, the Indian Ocean, the southeastern Atlantic, and the tropical Pacific. It is seen relatively frequently in the eastern tropical Pacific, especially near Hawaii and off Japan. Specimens have been maintained alive in captivity in Hawaii and Japan.

In no area is the pygmy killer whale described as abundant. There is no fishery specifically for pygmy killer whales, although a few are sometimes among the small cetaceans driven ashore by Japanese fishermen. Also, some are killed inadvertently by eastern tropical Pacific yellowfin-tuna fishermen.

Seen from below, pygmy killer whales can be distinguished from other "blackfish" by their rounded flipper tips and white ventral markings. (Off southeastern Lanai, Hawaii: © Bill Curtsinger.)

CAN BE CONFUSED WITH: The pygmy killer whale resembles the false killer whale, but is much smaller and can be distinguished at close range by the zones of white coloration. False killer whales are almost all black on the dorsum (except for the sometimes lighter gray areas behind the head), and their maximum length is at least 5.5. m. Pygmy killer whales are dark gray on the back, are often lighter on the sides, and show a region of white on the belly that may extend high enough up onto the sides to be visible on a surfacing animal. Furthermore, their maximum length is only about 2.7 m.

The pygmy killer whale may also be confused with the melon-headed whale, which is similar in size and coloration. So little is known of the two species' appearance and behavior at sea that only experienced observers can distinguish them reliably. At close range the shape of the flippers (rounded in *Feresa,* pointed in *Peponocephala*) and of the head (more blunt and definitely beakless in *Feresa,* tapered with a faint but sharp beak in *Peponocephala*) should help separate these two look-alike blackfish. Also, the white areas on the flanks of some pygmy killer whales are generally absent on melon-headed whales.

With a specimen in hand, the separation is simplified. Melon-headed whales have 21 or more teeth in each row; all other blackfish have fewer than fifteen.

False Killer Whale

Pseudorca crassidens
(Owen, 1846)
DERIVATION: from the Greek
pseudos for "false," and the Latin
orca for "a kind of whale"; from
the Latin *crassus* for "thick,
stout," and *dens* for "tooth."

ZONES 1 TO 5

DISTINCTIVE FEATURES: Flippers narrow, tapered, bearing a distinct hump on leading edge; body slender, almost all black; head narrow, tapered, sometimes lighter on top and back; mouth long, curved at gape, bearing eight to eleven large, conspicuous teeth in each row; distribution, principally warm temperate to tropical.

DESCRIPTION: Males may reach 6.1 m, though in most areas maximum length is 5.5. m and maximum weight about 1.4 tons. Females are slightly smaller, rarely exceeding 4.9 m. Newborn are usually about 1.5 to 1.8 m.

The head is small relative to body size, and smoothly tapered from the area of the blowhole forward. The closed mouth appears long and arched upward from the tip of the lower jaw, which is usually well behind the overhanging upper jaw, to the gape.

The body is long and slender. The dorsal fin, located just behind the midpoint of the back, is tall (to 40 cm) and falcate, and can range from rounded to sharply pointed on the tip. The flippers have a broad hump on the front margin near the middle, a characteristic that is diagnostic for the species.

A herd of false killer whales churns the sea surface as they sprint beside a research vessel. (Eastern tropical Pacific, February 9, 1977: Robert L. Pitman.)

The body is all black except for a blaze of gray on the chest between the flippers, and an area of light gray that may be present on the sides of the head. This area is often accentuated in bright sunlight. The ventral blaze varies from barely visible to light grayish white, similar to but generally fainter than that on pilot whales.

There are eight to eleven large, conspicuous teeth, circular in cross section, in each jaw. These teeth are often visible in the open mouths of free-ranging animals.

NATURAL HISTORY: False killer whales are gregarious, often forming herds of more than a hundred individuals, and often associating with other cetaceans, such as bottlenose dolphins. These herds usually contain individuals of both sexes and all age groups, and appear to have strong social cohesion. They have been assumed to be primarily oceanic, rarely approaching land except near oceanic islands and land masses with deep water nearby. Records off Venezuela have been exclusively from coastal waters. Migrations have not been described, though the most northerly extensions of the range, at least in the northeastern Pacific, appear related to spring-summer warming of the waters.

Both males and females reach sexual maturity at lengths of 3.2 to 3.8 m. Calves have been seen throughout the year, indicating that there is probably no fixed breeding or calving season.

These animals are frequently seen playing in and riding the bow waves of vessels, have been seen riding ground swells, and often jump completely out of the water. False killer whales have been suc-

A high-speed false killer whale breathes amid a spray of its own making. (Red Sea, December 1982: Hal Whitehead.)

cessfully maintained in captivity in Japan, Hawaii, and California, where they have proven durable, tractable, and highly trainable. Their larger size notwithstanding, they are capable of high leaps that rival those of smaller dolphins in the wild; their often high-speed swimming motions may be interrupted by rapid turns, acceleration, and sudden stops, particularly during feeding periods.

False killer whales feed primarily on squid and large fish, and have acquired a bad reputation in some quarters because they steal fish from the lines of fishermen, both commercial and sport. They have been seen eating dolphin fish (mahi-mahi) off Hawaii, and are one of several "blackfish" that more and more fishermen report seeing attack porpoises escaping seine nets in the tropical eastern Pacific. Large herds of false killer whales sometimes strand themselves dramatically on a beach, where their mass death invariably causes a flurry of excitement among scientists and the public.

DISTRIBUTION AND CURRENT STATUS: Though current evidence suggests they are nowhere abundant, false killer whales are widely distributed in warm temperate and tropical waters around the world. There are occasional records of vagrants in northern temperate waters. They are found off the Atlantic coast of North America from Maryland south, in the Gulf of Mexico, near Cuba and the Lesser Antilles, and in the Caribbean Sea as far south as Venezuela. They continue to be seen down at least to northern Argentina. On the opposite side of the Atlantic, they have been recorded stranded as far north as the central Norwegian coast and routinely range to the northern coasts of the British Isles in summer. From those poorly defined northern limits they range south past the tip of South Africa. They are known also from the Mediterranean Sea.

In the Pacific this species has been reported in Prince William

Sound, Alaska, and near 45° N in the northwestern Pacific, though sightings north of southern California (in the east) and northern Japan (in the west) are considered unusual. They have also been observed in Hawaii, though relatively infrequently, and in the South and East China Seas. In the southern hemisphere they have been reported as far south as New Zealand and Peru. They have been sighted throughout much of the Indian Ocean, though details of distribution there are not known.

False killer whales are occasionally taken for food in mass shore drives off Japan, where fishermen have also killed them in order to reduce perceived competition for prized fish resources. Dozens of specimens per year are taken accidentally in tuna long-line fisheries, and mass strandings, occurring sporadically around the world, may each account for hundreds of deaths.

CAN BE CONFUSED WITH: The false killer whale can be confused with the killer whale, the pilot whale, or the smaller, poorly known pygmy killer and melon-headed whales. It can be distinguished from the killer and pilot whales in the following ways.

The killer whale has a tall, unmistakable dorsal fin, and striking regions of white coloration; its body is much chunkier, its maximum length much greater (more than 9 m), and its head much broader and more rounded. Though basically all black (except for variable ventral gray and a post-dorsal saddle that is sometimes present), pilot whales have a rounded bulbous head, which pushes a broad crescent of white water as the animals surface, and a thick broad dorsal fin, low in profile, located very far forward on the back.

When specimens are in hand, the killer whale and pilot whale can be readily distinguished from the false killer whale by other features. The killer whale's flippers are huge and paddle-shaped, the ventrum all white. The pilot whale's flippers are long, slim, and acutely pointed. Both species' flippers lack the hump on the forward margin that is characteristic of the false killer whale.

At sea the false killer whale can be distinguished from the pygmy killer whale and the melon-headed whale primarily by its larger size and differences in coloration. The false killer whale grows to 5.5 m in length; pygmy killer and melon-headed whales reach only 2.4 to 2.7 m. The pygmy killer whale has an extensive region of white on the belly, which may extend onto the sides, and both pygmy killer and melon-headed whales have a white region on the lips, usually lacking or indistinct on the false killer whale.

Killer Whale

Orcinus orca
(Linnaeus, 1758)
DERIVATION: *Orcinus* may be
from the Latin *orca* for "a kind
of whale," or from the Latin
orcynus for "a kind of tunny,"
referring to the species'
resemblance to tuna or to its
habit of preying on them.

ALL ZONES

DISTINCTIVE FEATURES: Tall, wide dorsal fin, falcate in females, triangular in adult males; striking black and white pigmentation pattern (white patch above and behind eye, extensive white ventral patch extending onto flanks, and gray saddle behind dorsal fin); conical head; broad, paddle-shaped flippers.

DESCRIPTION: Males grow to lengths of at least 9.5 m and weights of eight tons or more. Females rarely exceed 7 m and four tons; they are less robust in overall body shape than males. At birth killer whales are 2.1 to 2.4 m long, and weigh about 180 kg. The body tapers anteriorly, and the snout is probably best described as conical, with a very indistinct beak and straight mouthline.

By far the most conspicuous external feature is the dorsal fin, which reaches exaggerated proportions in adult males. Although the dorsal fin of females and immature males is falcate and almost dolphinlike in appearance, that of large males approaches the shape of a very tall (to 1.8 m) isosceles triangle; sometimes it even cants forward slightly. In both sexes the dorsal fin is tall enough to be seen at a great distance.

The dorsal fin of a large male killer whale, with the scenic Royal Society Range in the background. (McMurdo Sound, Antarctica, Jan. 1981: S. Leatherwood.)

The flippers of killer whales are also impressive appendages. When an animal breaches or spyhops, which happens fairly frequently, these enormous rounded appendages are evident. The flukes have a slightly concave trailing edge and a distinct median notch.

The coloration of killer whales is handsome and distinctive, made up mainly of well-defined areas of shiny black and cream or white. In general, the dorsal half of the body is black, interrupted only by a light gray patch behind the dorsal fin. This saddle is believed to differ for each individual whale, and the shape and intensity of the saddle probably differs consistently between geographical races or stocks. An excellent field mark for identification at sea is the oval, white eyepatch, situated behind and above each eye. The chin, throat, and undersides of the flukes are white, and the white of the throat continues along the ventral midline, narrowing between the flippers. This belly patch branches into three segments posterior to the navel, one on either flank, and one continuing past the anus. As with the saddle, the shape of these color elements can be used to identify geographical populations of killer whales. There are usually about ten to twelve large conical teeth in each row; they curve toward the throat.

NATURAL HISTORY: Killer whales appear to be moderately gregarious animals, with strong social bonds and stable group structure. The size of their social units, usually called pods, ranges from a few individuals up to about 30. Sightings of herds of a hundred or more probably result from short-term coalitions of several pods. Pods can contain adults of both sexes, as well as calves and juveniles.

The reproductive characteristics of killer whales have not been firmly established. Gestation lasts for at least a year, perhaps several months longer. Calves are born mainly in fall and winter, and they appear to remain dependent for more than a year. The calving interval is thought to average more than two years.

Seasonal movements seem to be conditioned in polar regions by ice cover and in other areas primarily by availability of food. In much of their range killer-whale pods seem to be year-round residents. Major pinniped rookeries, many of which are occupied seasonally, attract marauding pods of killer whales while seals are present.

The frequent likening of killer whales to wolves is apt, since both species act as top predators, maintain complex social relationships, and hunt mammalian (and other) prey in a coordinated, cooperative fashion. There seems to be no palatable marine organism of any size that is safe from attack. Virtually all oceanic cetaceans and pinnipeds, penguins and other seabirds, sea turtles (leatherbacks, at least), many kinds of fish (especially herring and salmon), and even their own kind are eaten at times by killer whales.

Captivity has dramatically altered the status of killer whales. In the past they were viewed only as dangerous competitors for marine resources, and the standard response to their appearance was fear and hostility. Today in many parts of the world they are the object of affection and fascination. Killer whales survive for long periods in captivity, and they are readily trained to perform awe-inspiring antics.

DISTRIBUTION AND CURRENT STATUS: Killer whales are genuinely cosmopolitan, occurring among ice floes in polar latitudes and in equatorial regions. The only limits to their distribution seem to be ice cover, shortage of prey, and human predation. Though they are said to be most abundant within 800 km of continental coasts, they should be looked for in all oceans and major seas at any time of year. Local populations in protected waters of Washington State, British Columbia, and southeastern Alaska have been studied intensively with nonlethal techniques.

Commercial whalers throughout the world have captured killer whales fairly regularly, but no fishery primarily aimed at killer whales has developed. Small-whale hunters from Norway processed more than 1,400 taken in the northern North Atlantic between 1938 and 1967. Large numbers were taken for the first time in the southern hemisphere by Soviet commercial catcher boats during the 1979–1980 whaling season. In addition to commercial hunting, killer whales have been subjected to control measures because of concern about their depredations on herring, and sixty-seven killer whales were captured live in Washington and British Columbia for the aquarium industry between 1962 and 1973. The center of such activity has shifted in recent years to Icelandic waters, where these whales are resented for their interference with commercial fisheries.

CAN BE CONFUSED WITH: Their unique pigmentation and distinctive dorsal fin should make killer whales easy to recognize at close range. When females and young are seen at some distance, and are not accompanied by an adult male with a very tall dorsal fin, they might be confused with false killer whales and Risso's dolphins, neither of which is likely to venture into polar regions, but both of which inhabit the temperate and tropical portions of the killer whale's range.

False killer whales are somewhat smaller and more slender of build, and lack the striking white markings of the killer whale. Their head is tapered, but entirely beakless and rounded at the snout. Risso's dolphins are smaller still, and adults appear all or partly white on the back and head, with many scratches and scars. The rounded, beakless snout does not resemble the killer whale's in any way. Both the false killer whale and Risso's dolphin have tapered, sickle-shaped flippers, in contrast to those of the killer whale, which are broad and spatulate.

The bulbous head and broad-based, low-profile dorsal fin of the pilot whale should make it easy to distinguish from the killer whale. Nevertheless, the two are often confused by untrained observers.

Puget Sound is killer-whale country; sometimes these animals are seen in the shadow of Seattle's skyline. (Seattle, Washington, August 1977: K. C. Balcomb.)

Long-finned Pilot Whale

Globicephala melaena
(Traill, 1809)
DERIVATION: from the Latin
globus for "globe, ball," and the
Greek *kephale* for "head"; from
the Greek *melanus* for "black."

ZONES 2, 3, 4, 5, 8,
AND POSSIBLY 1

DISTINCTIVE FEATURES: Bulbous forehead ("pothead"), possibly exaggerated in adult males, and short, almost imperceptible beak; prominent, falcate dorsal fin, low in profile, with long base, located in front half of back; flippers long and sickle-shaped; slate gray to black, except for light markings on throat and belly and sometimes behind dorsal fin and eye; temperate to subpolar distribution in all oceans except North Pacific.

DESCRIPTION: Maximum size in males is about 6.2 m and 3 tons; in females, 5.4 m and 2 to 2.5 tons. Size at birth is about 1.75 m and 80 kg. The body has an elongated wedge shape when viewed dorsally. There is only the slightest hint of a beak below the high, rounded forehead, which may overhang the snout, particularly in old males. The mouthline is upcurved. The crescentic blowhole is set slightly to the left of center on top of the head.

The flippers are long (one-fifth of body length or more), sickle-shaped, and acutely pointed on the tip. There is a thick keel on the tail. The flukes are nearly pointed at the tips, with a concave trailing edge and a deep median notch. The dorsal fin is one of the most distinctive external characteristics. It is low in profile, markedly longer at the base than at the peak, set far forward on the animal's back, and falcate to flaglike in appearance. In adult males the fin may have a thicker leading edge and rounder form than in females.

At sea this whale generally appears to be slate gray to black all over; when dry on the beach, dark chocolate. There is usually a large, anchor-shaped patch of grayish white on the throat, and a pale stripe along the midline of the underside that widens posteriorly. In the North Atlantic some larger animals have a gray saddle behind the dorsal fin, but this feature is more prevalent in the southern hemisphere. In some herds, there is also a more or less conspicuous gray streak behind the eye, extending a few centimeters upward and backward, in the shape of an elongated teardrop. In many long-finned pilot whales seen in the southern hemisphere, the blaze behind each eye and the saddle behind the dorsal fin are white. Young animals are generally paler than adults. There are eight to twelve pairs of peglike teeth in the upper and lower jaws.

NATURAL HISTORY: These are gregarious whales, sometimes forming aggregations of several hundred to more than a thousand, and not infrequently found with other, smaller odontocetes, primarily bottlenose dolphins. However, they are usually seen in smaller groups, of less than fifty. The cohesiveness of herds varies according to activity: they are tight while on the move or being chased, loose while feeding. Drive fisheries and mass strandings have facilitated the study of herd structure. No consistent pattern of age or sex segregation has been discerned; however, it appears that many fewer males than females survive to adulthood. Pilot whales are probably polygynous.

A long-finned pilot whale lunges through a wave to breathe, revealing a faint hint of a beak and a prominent white post-ocular blaze. (Near the Antarctic Convergence at 57° 41.5' N, 165° 44' E, December 20, 1980: F. Kasamatsu. Japan Whaling Association.)

Carcasses of pilot whales, stranded by unknown but presumably natural causes, litter a New Zealand beach. (Near Pakawan, Farewell-Spit, South Island, October 22, 1962: © Jan P. Strijbos.)

The peak seasons for mating are spring and early summer. Most calves are born during late summer after sixteen months of gestation; some calving occurs throughout the year. Nursing lasts for about twenty months, one of the longest periods of parental dependence known for any cetacean. Age at sexual maturity in females is about six years; in males, twelve. The calving interval probably averages three years.

In Newfoundland, where they have been most thoroughly studied, long-finned pilot whales arrive inshore in June or July, and remain until late October or November. The winter grounds of these herds are believed to lie offshore, in waters influenced by the Gulf Stream. Some pilot whales apparently spend the entire year in offshore waters. No well-defined migration has been mapped in other areas, although there is clearly some seasonal variation in distribution.

Pilot whales are generally methodical and unexpressive at the surface. Entire herds may be seen "resting" at the surface, a gathering of dark log-like rafts, with no apparent motion. They sometimes hang vertically with the head and forepart of the flippers exposed, i.e., spyhopping or pitch-poling. Lobtailing (slapping the surface with the tail) is common. The very infrequent breaching is usually done by younger animals. They do not ride bow waves. Various dolphins are sometimes present in or near pilot-whale herds. In captivity, pilot whales have proven to be quick learners and impressive performers.

By far the primary food item is squid, and it is presumably the movements of these abundant invertebrates that determine the daily activity patterns and seasonal movements of pilot whales. Some cod and other fish are eaten when squid is lacking.

Predators are unknown. However, pilot whales tend to strand, both individually and in herds of several hundred, and this tendency may well be the most serious natural mortality factor. The life span has been estimated to be 40 to 50 years.

DISTRIBUTION AND CURRENT STATUS: This species is abundant and widely distributed in cold temperate waters of the North Atlantic and the southern hemisphere. The northern and southern populations are separated by a wide tropical belt and are sometimes recognized as subspecies (*edwardi* in the south, *melaena* in the north). Distribution in the North Atlantic is from West Greenland and Iceland south to Cape Hatteras, including the Gulf of St. Lawrence, and from the Barents Sea south to northwestern Africa, including the North, Baltic, and Mediterranean Seas. It is found not only near land masses, but continuously across the temperate North Atlantic. In the southern hemisphere these whales are found near all major land masses and in pelagic waters of the temperate and subantarctic regions, especially in the cold Humboldt, Falkland, and Benguela Currents, associated with the West Wind Drift. It is not clearly established whether their range includes the cold temperate North Pacific.

The species has been heavily exploited for meat and oil in the North Atlantic. The two largest-scale and best-known recent fisheries are those at the Faeroes and Newfoundland. In both areas the

The end of a successful drive in the centuries-old pilot-whale fishery at the Faeroes.

animals are sighted close to shore (usually in summer), and driven en masse into shallows, where they are shot, harpooned, or lanced. The Faeroese fishery is one of the most protracted and stable of its kind, with the catch for many centuries averaging a few hundred to more than a thousand per year. The Newfoundland fishery, organized to supply food for ranch mink, began in 1947 and closed in 1972. The catch reached close to 10,000 in 1956, and had declined to one or two hundred by the early 1970s. The Newfoundland stock of perhaps 60,000 was seriously depleted, but it now may be recovering. The Norwegian small-whale fishery catches this species; some are taken off West Greenland; and at least a few become fouled in fishing gear and die each year off North America. This species inhabited the North Pacific off Japan as late as the tenth century, but no recent specimens have been collected and identified.

CAN BE CONFUSED WITH: In the southern part of its North Atlantic range and in warm parts of its southern-hemisphere range, the long-finned pilot whale is likely to be confused with the false killer whale. The head of the latter is more tapered, and has no beak; the mouth is longer; and the dorsal fin is more slender, more erect, and situated farther back on the body. False killer whales are more expressive, occasionally riding bow waves and leaping clear of the water.

It is virtually impossible to distinguish the long-finned pilot whale from the short-finned pilot whale at sea, though one can guess because the long finned pilot whale's distribution is antitropical, and the short-finned pilot whale's is tropical. With stranded specimens the tooth count can be helpful (short-finned pilot whales generally have fewer: seven to nine versus eight to twelve), as can the ratio of flipper to body length, one-sixth versus one-fifth, but neither feature is consistently reliable. Museum preparation of the skull is often the only means of making a reliable judgment.

Short-finned Pilot Whale

Globicephala macrorhynchus
Gray, 1846
DERIVATION: from the Greek
makros for "long, large," and
rhynchos for "snout, beak."

ZONES 1 TO 5

DISTINCTIVE FEATURES: Similar to *G. melaena,* except flippers shorter and teeth fewer (seven to nine in each row); distribution generally more tropical.

DESCRIPTION: General appearance is very similar to that of the long-finned pilot whale. Maximum known length in males is about 5.4 m; females, 4 m. These figures, as well as the length at sexual maturity, may differ regionally. Males attain sexual maturity at 4.2 to 4.8 m, but probably do not reproduce successfully until at least 5 m. Females mature at 3 to 3.3 m. Length at birth is about 1.4 m.

The head is thick and bulbous, a characteristic that reaches its extreme in the flattened or squarish appearance of the front of the head in mature males. Flippers are proportionately shorter than those of *G. melaena,* one-sixth or less of body length. The gray ventral markings are less vivid and extensive on this species, but the postdorsal saddle is often large and conspicuous. The lightly pigmented blaze behind the eye is present but often muted. A light chevron may be present behind the blowhole on the neck of this species. There are seven to nine peglike teeth in each side of the upper and lower jaws.

Aquarium shows provide contact with cetaceans for millions of people who will never go to sea. Here a captive short-finned pilot whale performs. (Sea World, San Diego, California, July 1982: Jerry Roberts, Sea World, Inc.)

NATURAL HISTORY: In most respects this species is similar to the more antitropical long-finned pilot whale; however, there have not been any detailed studies of its biology comparable to those made in association with drive fisheries for long-finned pilot whales. Efforts off southern California toward live capture for research and public display have allowed some study of the short-finned pilot whale's behavior and physiology, and mass strandings in many areas have made specimens available.

Groups of a few to several hundred are common, often in company with dolphins (especially bottlenose dolphins). The bottlenose dolphins and male pilot whales tend to remain on the outer edge of the herd. Though they may be found fairly close to shore anytime, short-finned pilot whales appear inshore off southern California in early spring, coincident with the appearance of spawning squid. Seasonal catching peaks off Japan indicate that a similar inshore-offshore pattern may exist there. Outside squid season, the

Underwater, the short-finned pilot whale's light markings become accentuated. (Off Kealakekua Bay, Hawaii, March, mid-1970s: © Bill Curtsinger.)

whales usually are found offshore. Young are apparently born throughout the year, as might be expected in a tropical species not subject to climatic extremes. Radiotelemetric studies have shown that these whales can dive to depths of at least 610 m.

DISTRIBUTION AND CURRENT STATUS: The short-finned pilot whale is cosmopolitan in tropical and warm temperate waters. It

Short-finned pilot whales, known locally as blackfish, are hunted for food and oil in a West Indian shore fishery, employing, in part, pre-modern whaling technology. (West of Barrouallie, St. Vincent: R. V. Walker.)

has been recorded as far north as Delaware Bay and France, but generally ranges no farther north than Bermuda and Cape Hatteras (possibly Virginia in summer), south to Venezuela, and from Madeira and northwestern Africa south to perhaps Senegal. It is common in the Caribbean Sea and Gulf of Mexico. The species' distribution in the southern hemisphere is poorly known. It probably inhabits most of the warm waters of the Indian Ocean and along the coasts of northern Australia.

In the North Pacific, taxonomic problems cloud the picture. Pilot whales have been reported as far north as the Gulf of Alaska, but are much more common from Point Conception (central California) south to Peru and from Japan southward. They are also known from around Hawaii and the eastern tropical Pacific.

Pilot whales were frequently killed in the tropics, especially off northwestern Africa and in the West Indies, by nineteenth-century pelagic whaling crews for practice (their main target being sperm whales) and to obtain watch oil and meat. Small hand-harpoon fisheries in the West Indies began in the 1930s, and continue to account for several hundred whales per year. Drives have been conducted sporadically in Japan and Okinawa, and about a hundred pilot whales per year are still taken by harpoon gun in Japan. Some die in the purse-seine fishery for tuna in the eastern tropical Pacific, where the total population of pilot whales has been estimated as 60,000. The localized and irregular exploitation of short-finned pilot whales seems to have allowed the species to remain abundant and widely distributed. Short-finned pilot whales adjust well to captivity; so some live capture has taken place, particularly off southern California.

CAN BE CONFUSED WITH: Because their ranges are very similar, the short-finned pilot whale is likely to be confused with the false killer whale, certainly with the long-finned pilot whale where their ranges overlap, and in tropical portions of its range, perhaps with the pygmy killer whale and the melon-headed whale. Both of the latter have a more recurved, slender, pointed dorsal fin and a longer, more tapered head.

Rough-toothed Dolphin

Steno bredanensis
(Lesson, 1828)
DERIVATION: *Steno* from the
Greek *stenos* for "narrow"; Van
Breda was an artist who painted
the type specimen.

ZONES 1 TO 5

DISTINCTIVE FEATURES: Body roughly cone-shaped in front of the
flippers; beak long and slender, with no crease separating it from the
forehead, as there is in most dolphins; body often covered with
yellowish white blotches; lips and tip of snout white; eyes dark; distribution primarily pelagic, in tropical to warm temperate regions.

DESCRIPTION: Maximum reported length is close to 2.8 m. Females
are sexually mature at least by the time they attain a length of 2.35
m, and some data suggest both males and females are sexually mature by 1.8 m. Size at birth is unknown.

The shape of head and snout is one of the species' most distinctive features. The forehead (melon) slopes gently onto a long and
slender beak, slightly compressed laterally and lacking the transverse crease that separates the melon from the snout in other
beaked dolphins. Because both the forehead and the sides of the
head slope smoothly, the entire body in front of the flippers appears
very long and nearly conical in both dorsal and lateral profile.

The body generally is slim, although larger individuals may be
robust, much like large bottlenose dolphins. The flippers are rela-

tively long (one-seventh of body length) and the flukes wide (reaching one-fourth of body length). The moderately tall, falcate dorsal fin, slightly more prominent than that of a comparable-size bottlenose dolphin, is situated near midbody.

Though varying in color, rough-toothed dolphins are usually dark gray to purplish black on the back, with yellowish white or pink blotches on the sides of the body. The ventral surface is white, sometimes with a pinkish cast. The flippers and flukes and a region around the eyes are dark. The lips and often a large surrounding region of the lower jaw are white or at least white-flecked. Individuals

A synchronous blow by a pair of rough-toothed dolphins. (Off Hawaii, January 1980: Bernd Würsig.)

frequently have white streaks or scars on various parts of the body.

Rough-toothed dolphins have been reported to have 20 to 27 teeth in each row. As the common name implies, these teeth are not smoothly conical like those of other dolphins; rather, the crown is usually marked by a series of fine vertical wrinkles.

NATURAL HISTORY: The species is poorly known from the wild. Herds of as many as several hundred animals have been observed, though groups of fifty or fewer are more common. They are seen from time to time with pilot whales and bottlenose dolphins, much less frequently with spotted and spinner dolphins. Some have been found associating with schools of yellowfin tuna in the eastern tropical Pacific. They are not especially wary, but show less propensity to bow-ride than most species of wild dolphins. Recently small groups have been noted to skim, swimming rapidly with the snout continually near the surface and the dorsal fin continuously exposed, for protracted periods. Pelagic octopus, squid, and remains of several species of fish have been found in stomachs of examined specimens. Individual and mass strandings occur occasionally.

Rough-toothed dolphins have been maintained in captivity in Japan and Hawaii, where a long-lived individual still performs. An apparently healthy calf, the result of a female *Steno* mating with a male *Tursiops*, survived for almost five years at Sea Life Park in Hawaii. Unfortunately, it died before its ability to reproduce could be tested.

DISTRIBUTION AND CURRENT STATUS: Rough-toothed dolphins appear to be widely distributed, but are not known to be numerous in any specific area. They occur in tropical and warm temperate seas around the world, especially far offshore in deep water. Small numbers have been reported from widespread locations in the Indian (including the Gulf of Aden and Mossel Bay, South Africa, to much of the Indo-Australian Archipelago), Atlantic (Netherlands to the Ivory Coast, Virginia to the West Indies), and Pacific (northern California to Peru, northern Japan to New Zealand) Oceans. They wander infrequently into the Mediterranean Sea. Animals stranded in cold temperate waters are considered vagrants. When live, free-ranging rough-toothed dolphins have been reported, associated sea-surface temperatures have always been above 25° C.

Separate stocks have been suggested for the Atlantic and Indo-Pacific because of differences in rostrum lengths, but data are far from conclusive. Rough-toothed dolphins are reportedly rare off Japan and in the heavily studied eastern tropical Pacific. Small numbers are taken for food off the Izu Peninsula, Japan, in harpoon (and rarely in drive) fisheries. Smaller numbers die annually when they become entangled in seine nets of tuna fishermen who are catching yellowfin and skipjack tuna in the eastern Pacific.

CAN BE CONFUSED WITH: In most of their tropical offshore range, rough-toothed dolphins could be mistaken for bottlenose dolphins, spotted dolphins, and spinner dolphins. Unlike rough-toothed dolphins, these latter three species have a distinct crease or apex separating the forehead (melon) from the beak. The bottlenose dolphin's beak is shorter and stubbier, and its coloration more uniform, usually without the blotches and white lips. Spotted dolphins, with an intermediate-length beak, have spots on the body, and these, though they differ regionally, increase in number and intensity with age; furthermore, their color pattern in general is more complex, particularly in the cape and on the face. Spinner dolphins have a tall, sometimes erect, even forward-canting dorsal fin, and a complex color pattern, ranging from mostly gray with a light belly to handsomely marked with three color zones. Most have black lips and a black-tipped beak. The blotched coloration,

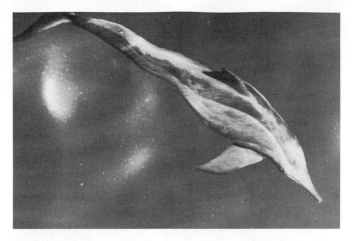

The rough-toothed dolphin's cape is a thin band of varying width along the midline of the back and tail stock. (Off Hawaii, January 1980: Bernd Würsig.)

very narrow cape (dissipating to a thin line in front of the dorsal fin), and white lips of the rough-toothed dolphin should help distinguish it from most of the other beaked dolphins.

In portions of their range where they may be sympatric with hump-backed dolphins, rough-toothed dolphins can be readily distinguished from immature hump-backed dolphins by the features described in the preceding paragraph. If mature hump-backed dolphins are present, the hump that occurs at the base of the dorsal fin and prompts the common name will make confusion unlikely. Adult hump-backed dolphins found east of Indonesia, however, apparently do not have the conspicuous humped dorsal fin, and thus may be difficult to distinguish from rough-toothed dolphins.

Tucuxi

Sotalia fluviatilis
(Gervais, 1853)
DERIVATION: *Sotalia* is believed
to be a coined name; from the
Latin *fluviatilis* for "of a river."

ZONES 2 AND 5

DISTINCTIVE FEATURES: Similar to *Tursiops truncatus*, except for smaller, nearly triangular dorsal fin, and tooth count, having 26–35 pairs in each jaw; limited to rivers and flooded jungles, as well as nearshore marine waters of northeastern South America and eastern Central America.

DESCRIPTION: Maximum length is about 1.9 m. Those inhabiting the Amazon may be smaller. The largest Amazon specimen reliably measured was about 1.6 m long, and females from that area are sexually mature at about 130 cm. Animals found in bays along the coast of Colombia appear to mature by at least 1.7 m. Length at birth in the central Amazon is 70 to 80 cm.

The body is similar to that of the bottlenose dolphin, though the prominent beak is less clearly demarcated from the forehead, and the body as a whole is stubbier. The dorsal fin is lower in profile (110 to 127 mm high) and almost triangular, curving only slightly backward near the peak.

Color differs according to habitat, but in general the back is dark (ranging from blackish, through brown and dull lead gray, to pale bluish gray) and the belly light (ranging from white, to gray, to pink). The dorsal and ventral fields extend onto the face and flanks to various degrees. There is sometimes a brownish band that extends from the dark color of the back in front of the dorsal fin back toward, but not reaching, the anus. Individuals living in the turbid Amazon are generally lighter in color than those in somewhat clearer coastal waters, often having whitish blotches on the tip of the beak and dorsal fin. There are 26–35 pairs of teeth in each jaw. These are often raggedly arranged, particularly in the lower jaw.

NATURAL HISTORY: Groups of two to 25 are seen in the Amazon Basin and in bays along the Caribbean coast between Panama and Colombia. The basic social unit in the Amazon may consist of groups of six or seven individuals. The tightness of their swimming formation suggests that they have strong social ties. Solitary animals and groups of two or three are often encountered in the lakes and streams inhabited during flood periods. They share much of the boutu's range, particularly in fresh water.

Tucuxi in the central Amazon give birth mainly in May to August, that is, during periods of high water. Females in this portion of the species' range are sexually mature by a length of about 130 cm. The gestation period is about 10.2 months.

There is reason to think the tucuxi undertakes mainly local movements, remaining within a limited home range. Those individuals that live in rivers and streams apparently disperse at flood periods deep into the inundated forests, though they tend to remain in the main channels more than the sympatric boutu. The tucuxi seems able to move freely between fresh, brackish, and marine environments.

The tucuxi is not always shy, but because much of its habitat is extremely turbid, it is not easy to observe. The intense, rapidly repeated clicking sounds made by this dolphin probably allow it to echolocate in a visually opaque medium. Although animals in coastal waters supposedly show little of themselves when surfacing, those in rivers frequently expose head and trunk as they breathe. Timed dive intervals in the Amazon ranged from five to 85 seconds (with a mean of 33 seconds), and individuals leaped as much as 1.2 m clear of the water. They often breach and fall back into the water on their sides.

Food includes fish and crustaceans (prawns and crabs). In Surinam armored catfish appear to be the mainstay of its diet. The worn condition of the teeth, even in young tucuxi, may be caused by the hard shells of some of their prey. In the central Amazon, prey species are primarily schooling fish that are not as sensitive as the boutu's prey to changes in water levels. Differences in food species and responses to flooding seem related to differences between these two species in schooling, dispersal, and hunting behavior.

Mortality factors are unknown. The species has been main-

In captivity the tucuxi is easily trained to exaggerate its high-jumping capabilities. (Duisburg Zoo, Duisberg, Germany: P. Schulz, courtesy of W. Gewalt.)

tained in captivity for brief periods in the United States, Europe, and Brazil.

DISTRIBUTION AND CURRENT STATUS: Finding out the distribution of the tucuxi is complicated by the fact that no less than five nominal species have been described. Most taxonomists currently rec-

A tucuxi mother and calf, caught by a local fisherman in the Amazon basin. (Upper Trombetas River, Brazil, October 10, 1969; courtesy of Frederico Medem, via James G. Mead.)

ognize all as belonging to a single species, distributed in waters of northeastern South America, from Santos, Brazil (24° S), northward to Lake Maracaibo in northwestern Venezuela and up the Caribbean coast to Panama. A stranding in Trinidad-Tobago has been reported. Its range includes nearshore coastal waters, and large river systems and their tributaries. It is said to be common in the Bay of Rio de Janeiro. In the Orinoco of Venezuela, it is known as far upstream as Ciudad Bolivar, and in the Amazon drainage it ranges at least 2,500 km upstream to near the base of the Andes.

One or more of the upstream riverine populations may be effectively isolated, but nothing is known with certainty about whether stocks are discrete. The species is considered common off the Surinam coast, off the coast between Rio de Janeiro and Santos, and in some portions of the Amazon Basin.

The tucuxi is sometimes killed by accident in gill nets set at river mouths in Surinam, and at least occasionally in drift or gill nets in Brazil. Local people along the Amazon apparently attribute some superstitious and medicinal value to parts of the tucuxi, which are sold in markets. How damming and other developments in South American river systems may affect the species is a topic of concern.

CAN BE CONFUSED WITH: It is important to distinguish the tucuxi from the larger boutu, which shares much of its riverine habitat. The best distinguishing feature is the dorsal fin, which in the boutu is replaced by a low, triangular hump. The tucuxi's beak is shorter, and the forehead bulge less prominent. In addition, the boutu is a slower and less active swimmer, much less demonstrative at the surface, and normally less gregarious.

The much-larger bottlenose dolphin will also be seen in some of the coastal and estuarine habitat of the tucuxi. However, only young bottlenose dolphins will be about the size of an adult tucuxi. These can be distinguished by their taller, broader-based, and more falcate dorsal fin, and, in stranded specimens, by their lower tooth count (18 to 26 pairs in each jaw).

Franciscanas also may be present in some portions of the tucuxi's range. At sea it is extremely difficult to tell these two species apart. Mature franciscanas have extremely long beaks, however, and certainly specimens in hand can be distinguished by the extreme difference in tooth counts and the rounded (rather than tapered) flippers of the franciscana.

Indo-Pacific Hump-backed Dolphin

Atlantic Hump-backed Dolphin

Sousa chinensis
(Osbeck, 1765)
DERIVATION: The derivation of *Sousa* could not be discovered; *chinensis* refers to China.

ATLANTIC HUMP-BACKED DOLPHIN

Sousa teuszii
(Kukenthal, 1892)
DERIVATION: Not known.

ZONES 3 AND 5
ZONES 2 AND 5

DISTINCTIVE FEATURES: Dorsal fin at midback consists of a long, ridged base surmounted by a small triangular or falcate fin-tip in adults, west of Indonesia; east of Indonesia the hump is lacking, but the fin is more pronounced; many animals have a speckled appearance; found in coastal waters of West Africa (*teuszii*) and in Indo-Pacific (*chinensis*); there appear to be two groups of undescribed relationship in the Indo-Pacific, one west of Indonesia, one east and south of Indonesia.

DESCRIPTION: Maximum reported length is about 2.8 m; weight up to 285 kg. The long beak is well-defined, the mouthline straight. The head somewhat resembles that of the bottlenose dolphin, though the hump-backed dolphin's beak is relatively longer, and the melon not so pronounced.

There is often a thick elevation, or hump, in the middle of the back of adults, and the small falcate or triangular dorsal fin appears to rest on this raised platform. The dorsal fin's profile is an unmistakable key to identification of those animals that possess a conspicuous hump (some individuals have a "normal" falcate dorsal fin). It appears that Atlantic hump-backed dolphins and those Indo-Pacific hump-backed dolphins found approximately to the west of Indonesia have the hump, but those to the east do not. When present, the hump is particularly evident as the animal arches its back to sound. The flippers are broad near the base, and bluntly tapered distally. The flukes are moderately concave along the trailing edge, with a distinct notch between the flukes; the flukes are frequently exposed as the animal dives.

Coloration can differ greatly, but is basically gray. An animal in the Persian Gulf was described as light gray, with dark bluish-gray

An adult Atlantic humpbacked dolphin fishes within a few meters of the beach as a photographer looks on. (Near Cap Timiris, Mauritania: courtesy of René Guy Busnel.)

spots "running parallel to the body axis that gave the impression of spotted stripes." Several specimens off Malabar were described as "steel gray" with "elongated black spots of a light bluish shade on their sides." The lower side and ventral surfaces were creamy white. Those off India are apparently solid lead gray; young animals in much of the eastern part of the species' range are cream-colored. The lower jaw is often cream-colored.

Off South Africa, where hump-backed dolphins have been studied in some detail, young calves are pale gray to off-white, deepening to grayish early in life. The skin remains an even and unblemished gray, sometimes with a purplish cast, until adulthood. In adults the dorsal fin and areas around its base, as well as the tip of the rostrum and sometimes the flukes, whiten considerably. Also, the skin, particularly that on the back half of the body, may be more

As this hump-backed dolphin surfaces to breathe, there can be little doubt about the origin of the species' common name. (In Salalah Harbor, Oman, January 10, 1982: Abigail Alling–WWF.)

or less scarred. In at least some portions of the South China Sea and the waters of Malaysia and Indonesia, adults are lighter, sometimes even snowy white, but calves are darker, steel-blue gray. In all regions, the belly is apparently lighter than the back and sides. Because of the extreme variability in appearance, it is best to consider the lumping of Indo-Pacific forms under *S. chinensis* to be tentative, and perhaps even arbitrary. There appear to be two morphologically different groups, one east and south, one west, of Indonesia.

There are no major external morphological differences, so far as is known, between *S. chinensis,* as found off East Africa and eastward approximately to Indonesia, and *S. teuszii*. However, the two groups are thought to be isolated geographically. Also, in specimens assigned to *S. chinensis* in the literature, there have been 29-38 teeth in each row, totaling more than 120, but most *S. teuszii* apparently have 26-31 per row.

NATURAL HISTORY: Most of what is known about hump-backed dolphins comes from experience with them in the Indian Ocean. These dolphins are often found singly or in pairs, though groups of twelve to twenty are not unusual. The average group off South Africa contains about six individuals. There is no obvious segregation by sex or age within populations. Social bonds do not appear to be particularly strong, with individuals moving indiscriminately among different groups. There is some evidence, however, that such small groups are part of larger coherent "schools" that maintain their identities for long periods of time. Hump-backed dolphins frequently are seen near bottlenose dolphins and finless porpoises.

Life-history information is sparse. In Plettenberg Bay, South Africa, mating and births occur throughout the year, but there is a strong peak of births in summer months.

They sometimes turn on their sides, waving a flipper in the air; raise their heads above water as if making a visual inspection of their surroundings; and leap clear of the water (especially young ones). There is much body contact and frequent aerial display. Frequently, adults swim in rapid circles. They do not seem perturbed by normal fishing operations; in fact, they sometimes prey on fish that fall out of working trawl nets. However, they do not bow-ride, and can be approached on a powered vessel only with difficulty. When not disturbed, hump-backed dolphins swim slowly, surfacing about every minute. When doing so they usually ascend steeply to the surface, often exposing the snout and head, and then flex far forward, accentuating the hump as they sound.

The preferred habitat of the Indo-Pacific hump-backed dolphin is shallow nearshore or inshore waters. Off South Africa they are rarely found more than 1 km offshore. Coasts with mangrove swamps and river mouths are favored, as are the edges of rock reefs, sandbanks, and mudbanks. Though it moves into rivers at times, it is definitely a saltwater animal, and seldom ventures beyond tidal influence. Seasonal movements are not well-documented.

Fish (for example, mullet and clupeids) are the only known prey. One of the most intriguing stories about cetacean interactions with people has been told of the Atlantic hump-backed dolphin. It is known to participate in cooperative shore seining for mullet by Mauritanian fishermen. When a school of mullet is sighted, the men beat the water with a stick in a characteristic manner. The procedure regularly appears to attract dolphins, mainly *Tursiops* and *Sousa*, which drive the mullet toward shore and into the fishermen's nets. However inadvertent this symbiotic relationship may be, it seems to be a well-established tradition at Cap Timiris, south of Cap Blanc. Those who have watched the men and dolphins fish together report that the hump-backed dolphins involved are usually mixed in schools of *Tursiops*. Both species come very near shore.

The diet of the Atlantic species has been the subject of some confusion since Kukenthal, who first described this dolphin, reported it to be herbivorous. Apparently the vegetable matter he found in stomach contents was ingested accidentally or was in the stomachs of fish which the dolphins had eaten, or he may have confused the stomach contents of a *Sousa* with those of a West African manatee, which inhabits some of the same areas as the Atlantic hump-backed dolphin. Graham Ross, who has studied hump-backed dolphins off South Africa, also attributed algae remains he found in stomachs to fish eaten by the dolphins.

DISTRIBUTION AND CURRENT STATUS: The Indo-Pacific hump-backed dolphin is widely distributed in coastal and inshore waters of the Indian and western Pacific Oceans. It prefers warm temperate and tropical latitudes. Its presence has been confirmed from the southern tip of Africa northward along the east coast of the continent to the Suez Canal; in the Arabian Sea and Persian Gulf; along the Indian subcontinent; throughout much of Indonesia; in Australian coastal waters, from the middle of the west coast northward, eastward, and southward to Sydney on the east coast; in New Guinea; and from Borneo northward along the Indo-Chinese coast to the Canton River. No records exist for the Philippines, but it may also be present there. Distribution probably is not continuous throughout this range, and the taxonomic status of various subpopulations remains unclear.

Some direct hunting has taken place in the Arabian Sea, the Red

Hump-backed dolphins (top) of both forms generally have a longer beak and more teeth than the similar bottlenose dolphin (bottom). (Sea World of Australia, Queensland: Peter Doggett.)

Sea, and the Persian Gulf, though it has reportedly decreased as the region's economic status has improved. *Sousa* are sometimes accidentally killed in antishark nets off Queensland, Australia, and Natal, South Africa, and in fishing nets off Pakistan.

The known range of the Atlantic hump-backed dolphin extends from the coast of Mauritania in northwestern Africa south to Cameroon and possibly to Angola. Since few studies have been done on cetaceans in West Africa, the range of this species has been only roughly mapped. It is not known to be common anywhere except in coastal waters of southern Senegal and off Cap d'Arguin in northwestern Mauritania.

At Cap Timiris the natives are fiercely protective of the dolphins, on whom they depend for assistance in fishing. However, they do not hesitate to utilize stranded animals for food and oil. In the past, hump-backed dolphins are known to have been caught accidentally, along with West African manatees, in beach seines and shark nets in Senegal.

CAN BE CONFUSED WITH: Hump-backed dolphins are most likely to be confused with bottlenose dolphins, which share their entire range. West of Indonesia the distinctive dorsal fin of the hump-backed dolphin should allow identification under most circumstances. East of Indonesia, where hump-backed dolphins lack the pronounced hump and have, instead, a dorsal fin similar to that of the bottlenose dolphin, differences in surfacing behavior, snout and melon shape, and adult coloration may be used to distinguish them. Hump-backed dolphins make a high roll, often exposing the long snout and inconspicuous melon, then make a sharp rolling or humping motion. Bottlenose dolphins generally rise more nearly parallel to the water surface, rarely exposing snout and head, and submerge with less arching of the back. Further, in this region adult hump-backed dolphins are light gray to white in color. Bottlenose dolphins are generally darker, steel-blue gray to dark gray.

White-beaked Dolphin

Lagenorhynchus albirostris
(Gray, 1846)
DERIVATION: from the Greek
lagenos for "bottle, flask," and
rhynchos for "beak, snout"; from
the Latin *albus* for "white," and
rostrum for "beak, snout."　　　ZONES 2 AND 8

DISTINCTIVE FEATURES: Short, thick beak, white or light gray on European side of range, sometimes dark off North America; two white or gray areas on each side of body, one in front of, and the other behind and below, the dorsal fin; posterior light area continues onto dorsal aspect of caudal peduncle; limited to northern North Atlantic.

DESCRIPTION: This dolphin is very similar to the Atlantic white-sided dolphin, although it is a bit larger and more robust. Maximum length is about 3.1 m. Newborn are about 1.2 m long. The tall, back-curved dorsal fin is located at midbody. The forehead is not bluff; the beak is short (5.1 cm long in adults), thick, and rounded. The flippers are broad at the base, curving backward and narrowing to a point. The caudal peduncle is thickened, giving it a strong keel above and below. The tail has a concave trailing edge and a shallow median notch.

　　　Coloration is distinctive. The back and sides are basically black or dark gray; the belly is white or light gray. The appendages are dark. The beak, as the common name implies, is often light gray to white, at least in European waters. In the western North Atlantic it may not always be white. There may be dark flecks behind the eyes, and a dark gray stripe connects the corner of the mouth with the forward base of the flipper. The best keys for recognizing this species while it is swimming are the two areas of pale coloration on the sides and flanks. The forward area begins above the flippers, and continues back along the side to the dorsal fin. The other originates below and behind the dorsal fin, and ends midway along the caudal peduncle, sometimes continuing across the dorsal ridge to meet the corresponding light area on the opposite side. There are 22–28 small

An acrobatic, upside-down jump by a white-beaked dolphin. (Off Labrador, 54° 32' N, 56° 07' W, August 16, 1982: Abigail Alling – WWF.)

teeth on each side of both jaws.

NATURAL HISTORY: White-beaked dolphins sometimes occur in herds of several hundred, especially in the eastern part of their range. In the west, smaller schools are more common.

Mating is believed to occur in the warmer months, and young are probably born between June and September.

Migrations are poorly understood. These dolphins are found near the northern limits of their range between spring and late autumn, apparently wintering to the south, where they may linger until late spring or even early summer.

The diet of the white-beaked dolphin includes squid, octopus, cod, herring, haddock, capelin, and sometimes benthic crustaceans.

Nothing is known about predation upon this dolphin. Strandings, usually of single, old individuals, are not uncommon in temperate parts of its range.

Obviously, some white-beaked dolphins in the western sector of their range do have a white beak. (Off Labrador, 54° 32' N, 56° 07' E, August 16, 1982: Abigail Alling – WWF.)

DISTRIBUTION AND CURRENT STATUS: This species is found in the northern North Atlantic, in cold temperate and subarctic waters. In the west it is common off Cape Cod during spring, and abundant at least seasonally off southern and western Greenland, Newfoundland, and Labrador, and throughout Davis Strait. In the east it is most common in the North Sea and around the Faeroes, but also present seasonally in the Norwegian Sea along the coast of Norway and in the southern Barents Sea east to Varanger Fjord and possibly Murmansk. It is abundant off southwestern Sweden, and numerous strandings have been reported for the Dutch and British coasts. It is said to be the most common dolphin in the waters around Iceland.

The white-beaked dolphin is not known to have been hunted on a major scale, although commercial catches supposedly have been made in Davis Strait and off Norway. Unfortunately, the record of its exploitation has not been well-documented. Some are accidentally killed in trawl nets in the North Sea.

CAN BE CONFUSED WITH: The white-beaked dolphin's range overlaps that of the Atlantic white-sided dolphin, and the two species are similar in appearance and behavior. The arrangement of the white or lightly pigmented patches on the flanks offers the best hope of telling them apart at sea. The white beak—when in fact it is white—of *L. albirostris* can also be a useful key.

Atlantic White-sided Dolphin

Lagenorhynchus acutus
(Gray, 1828)
DERIVATION: from the Latin
acutus for "sharp" or pointed."

ZONES 2 AND 8

DISTINCTIVE FEATURES: Narrow white patch on flanks; yellow or tan streak above white patch extending up toward ridge of tail; short, bicolored beak; limited to northern North Atlantic.

DESCRIPTION: Maximum length is about 2.7 m. Males are larger than females. At birth these dolphins are about 105 to 130 cm long, the average being approximately 120 cm. The forehead slopes at a low angle toward the beak, terminating at a deep crease that defines the beginning of a short, thick snout (to 6 cm long).

The pointed flippers are sickle-shaped and swept backward. The dorsal fin is tall, pointed, and distinctly back-curved; it is positioned at the middle of the back. The caudal peduncle has a thick keel above and below, and does not narrow laterally until very near the flukes, which have a strongly concave trailing edge with a shallow median notch.

Coloration is striking and distinctive. The upper jaw, back, flippers, and tail are black or gray; the belly white; the lower jaw light gray. The sides have zones of gray, tan or yellow, and white. A narrow dark stripe connects the anterior insertion of each flipper with the corners of the mouth, and there is a black eye patch, from which a thin line extends forward to the dark rostrum. There are two key field marks: one is an elongated oval patch of white or cream on each side, from below the dorsal fin to an area on the side above the anus; the other is a long tan or yellow swath on either side of the tail stock. These patches, clearly demarcated from the surrounding coloration, are frequently visible as the animal rolls at the surface. They are diagnostic for identification at sea.

The Atlantic white-sided dolphin has 29–40 sharp, pointed teeth on each side of both jaws.

NATURAL HISTORY: This species is sometimes observed in small groups of six or eight in inshore areas in summer, but offshore herds in particular often number many hundreds (more than 700 were once driven into a Norwegian fjord by fishermen). There is some evidence of segregation in these herds, with immature and newly matured animals largely absent from breeding groups.

Puberty may begin in males at between four and six years of age, although they may not participate in reproduction until fully mature. Females reach sexual maturity at five to eight years of age. Gestation lasts for about eleven months, with most calves being born in June and July. Lactation lasts for up to eighteen months. The calving interval is probably two or three years.

In the western North Atlantic, these pelagic dolphins appear to prefer 9° to 15° C waters over the continental slope between the Gulf Stream and the cold inshore water that originates with the Labrador Current. There may be a tendency to move inshore during warm periods, but no clear migratory pattern has been discerned.

This species is often wary of ships, but it does occasionally ride bow waves. It is frequently found associating with long-finned pilot whales and frolicking with feeding fin whales. Squid and herring are important parts of the Atlantic white-sided dolphin's diet, but it also eats smelt, silver hake, and several kinds of shrimp.

Mortality factors are not well-documented, although sharks and killer whales are probably at least occasional predators. Individual and mass strandings, sometimes of dozens of animals, are common. Atlantic white-sided dolphins are believed to live for as long as 27 years.

DISTRIBUTION AND CURRENT STATUS: This dolphin lives only in the northern North Atlantic. It is an offshore form that normally reaches no farther south than Cape Cod and the mid-Atlantic canyons (Hudson and Baltimore) off the United States in the west and the British Isles in the east. It is known from far inside the Gulf of St. Lawrence, and off West Greenland, Iceland, and Norway, and may occasionally penetrate as far north as the southern Barents Sea. It has been recorded in the Baltic and North Seas.

Large herds are most often encountered off Newfoundland and western Norway, where they historically have been killed, sometimes by the hundreds, in coastal drive fisheries. "Jumpers," as they are called in Newfoundland, were occasionally mixed in the catches there of pilot whales. There is no information on this species' status, but it is not now hunted on a major scale, and is considered regionally abundant. A catch of 44 animals was reported at the Faeroe Islands in 1976.

CAN BE CONFUSED WITH: Because of some overlap in distribution and similarity in build and appearance, the Atlantic white-sided dolphin and the white-beaked dolphin are likely to be confused. Careful attention to coloration helps one distinguish them at sea. The white-beaked dolphin has two pale or white patches on each flank, fore and aft of the dorsal fin, in contrast to the single, more centrally positioned white patch on each side of the Atlantic white-sided dolphin. The white ventral coloration of the white-beaked dolphin does not extend above the flippers, whereas it extends high onto the sides of the Atlantic white-sided dolphin. The presence of white on the dorsal aspect of the tail immediately behind the dorsal fin distinguishes the white-beaked from the Atlantic white-sided dolphin. In the eastern Atlantic, the white-beaked dolphin's all-white beak can be a useful mark of distinction, but it is not a reliable key on the western side. In dead animals tooth counts can be used to separate the two species. The Atlantic white-sided dolphin is often confused with the common dolphin, which also has a complex color pattern, including white and tan patches on the sides. The common dolphin's slimmer build, longer beak, and very distinctive dark saddle, defined by a V shape below the dorsal fin, should help to distinguish it.

Dusky Dolphin

Lagenorhynchus obscurus
(Gray, 1828)
DERIVATION: from the Latin
obscurus for "dark, indistinct."

ZONES 3 TO 5

DISTINCTIVE FEATURES: Lack of a prominent beak; erect, slightly hooked, two-tone dorsal fin; a pair of gray "suspenders" along the back; two dark, shadowlike blazes at midflank pointing tailward; found in temperate southern hemisphere.

DESCRIPTION: Maximum length for males is about 2.1 m. Most individuals are about 1.5 to 1.7 m long. The slight beak does not exceed a few centimeters. In some individuals only an indistinct crease and its dark coloration give definition to the end of the snout.

The moderately long flippers are distinctly curved on the anterior margin and taper to a blunt point. The dorsal fin is tall and exceptionally erect, only slightly recurved, and has a blunt peak.

The pigmentation pattern is complex. The entire dorsal surface is bluish black, as are the tail and snout. A pale gray band connecting the snout and eye sweeps down to the flipper in some animals. The dorsal darkness widens posteriorly as it approaches the dorsal fin, and extends onto the grayish flanks in the form of two prominent blazes that sweep ventrally and end adjacent or posterior to the dorsal fin, leaving a long, light-gray patch on the flanks. The countershading that results from this combination of gray and black gives the appearance of gray "suspenders" on the sides and back. The flippers and the back half of the dorsal fin are light gray. The belly and throat are white. There are 24–36 pairs of small teeth in the upper and lower jaws.

NATURAL HISTORY: Group size varies seasonally, at least in Cook Strait, New Zealand, and in Argentine waters. Hundreds are seen together in summer, but in winter schools of six to twenty are the rule. Groups of six to fifteen are most common in Argentine waters, but twenty to thirty of these small groups sometimes can be found within an area about 9 to 15 km in diameter.

In Argentina, calves are born mainly in summer. Preliminary results of studies in New Zealand indicate parturition occurs there in mid-winter (June to August). Gestation was estimated as eleven months; lactation eighteen months. The shortest sexually mature female measured 165 cm.

Major migratory patterns have not been demonstrated. Off Argentina dusky dolphins are found in some areas the year round. They seem to follow closely the movements of their primary prey (southern anchovies), coming inshore in spring and early summer, then moving offshore into deeper water in late summer and fall.

Dive times in Argentine waters depend on activity. At night the dolphins submerge for an average of only six seconds, traveling at about 5 km per hour. After dawn, as the search for and pursuit of food proceeds, they swim faster and dive longer. This pattern holds for spring and summer, but appears to be reversed in winter. In

Once it begins, a dusky dolphin may make 40 to 50 of these spectacular leaps, one after the other. (Off Kaikoura, New Zealand, February 15, 1981: Stephen Leatherwood.)

Swooping terns and a breaching companion probably signal that two dusky dolphins (left) have found food. (Off Kaikoura, New Zealand, February 15, 1981: Stephen Leatherwood.)

New Zealand waters dusky dolphins dive to known depths of at least 150 m.

This is an acrobatic species. If one in a herd begins to breach, others will often follow suit, causing a lively turmoil at the sea surface. Somersaulting is common, particularly after feeding episodes. Straightforward leaps, with splashless, headfirst re-entries, are typical when the dolphins are chasing fish. During actual feeding, noisy body slams seem to be used to herd prey and to call the attention of nearby dusky dolphin groups to the presence of food.

Dusky dolphins ride bow waves and are easy to approach. They are frequently seen playing with right whales off Argentina. Small numbers have been maintained in captivity for several years in New Zealand, South Africa, and Switzerland, and have proven amenable to training.

Dusky dolphins prey on fish, primarily southern anchovies off Argentina, and squid. However, their food habits are not well-known. Killer whales are known enemies, and are responsible for at least some of the dusky dolphins' movement inshore.

A multiple stranding involving six animals has been recorded.

DISTRIBUTION AND CURRENT STATUS: The dusky dolphin is an inshore species whose range is very nearly circumpolar in warm temperate and cold temperate waters of the southern hemisphere. It is most often seen off the Cape of Good Hope, Peninsula Valdes in Argentina, and New Zealand, where it is thought to be associated with the subtropical covergence. Its presence has also been established at Campbell Island, Kerguelen Island, and the Falkland Islands, in Magellan Strait, and off Chile and Peru. Though sometimes reported off mainland Australia and Tasmania, there are no verified records for these areas. Its distribution is apparently discontinuous; so there may be several geographically isolated stocks.

Dusky dolphins have been caught for human consumption in beach seines along Cape Peninsula, South Africa, and some are also taken in purse seines and with harpoons off South Africa. Hundreds are killed incidentally each year in fish nets off western South America and New Zealand. However, the species is not known to have been exploited commercially, and it remains abundant throughout most of its range.

CAN BE CONFUSED WITH: The dusky dolphin will readily be confused with its two southern-hemisphere congeners, the hourglass dolphin and Peale's dolphin. Pigmentation of the hourglass dolphin is much more sharply defined, with all-black appendages, and a neat black-and-white hourglass pattern on the flanks. The two white zones on each side, fore and aft of the dorsal fin, are conspicuous. Peale's dolphin can best be distinguished by its black face and throat, and the diagonal dark swath that cuts across each side at midbody. The dusky dolphin and Peale's dolphin have a more northern distribution, and, unlike the hourglass dolphin, are not likely to be seen south of the Antarctic Convergence. Peale's dolphin will be a source of confusion only around southern South America and near the Falklands, whereas the hourglass dolphin can be encountered almost anywhere in the more southerly portions of the dusky dolphin's range. The dusky dolphin also closely resembles the Pacific white-sided dolphin of the northern hemisphere.

Pacific White-sided Dolphin

Lagenorhynchus obliquidens
Gill, 1865
DERIVATION: from the Latin
obliquus for "slanting, sideways,"
and *dens* for "teeth."

ZONE 1

DISTINCTIVE FEATURES: Short, thick beak, clearly demarcated from forehead; dorsal fin sharply hooked, usually dark on leading edge, lighter posteriorly; back dark, but with two stripes or "suspenders" from head to near tail; distribution limited to North Pacific.

DESCRIPTION: This species reaches lengths of at least 2.3 m and weights of at least 150 kg. Both sexes reach sexual maturity by about 1.8 m. Maximum size and size at sexual maturity are believed to differ in different geographical regions. Length at birth has been reported to be about 80 to 95 cm.

There is a short but distinct beak, clearly demarcated from the forehead. The body is moderately robust on larger animals, but may be slimmer and more streamlined on others. The dorsal fin (the species' most distinctive feature in encounters at sea) is hooked, prompting a locally used common name, "hook-finned porpoise," and is usually black on the front, pale gray on the rear portions. The flippers are neatly curved, with a blunted tip, and are usually colored the same as the dorsal fin.

Normal coloration is black on the back, with striking light-gray sides and a pearl-white belly. The black color of the back is interrupted on each side of the dorsal fin by "suspenders," a pair of white stripes that begin in the gray color of the face, curve upward over the top of the head, extend past the dorsal fin to the area above the anus, and end as a widening zone on the sides of the tail stock. The light gray thoracic region is generally separated from the whiter belly by a black stripe from the gape to the all-black flippers, and from there to the black flanks. The beak and lips are black. The eyes are also dark, particularly as seen against the white face. Anoma-

Two Pacific white-sided dolphins hitch a ride on the bow wave of a research vessel, leaping and gamboling with ease before the ship's advance. (Off Baja California, 27° 40' N, 115° 42' W, March 9, 1979: Robert L. Pitman.)

lously colored animals do occur, apparently quite frequently in the northwestern Pacific. Specimens so far examined have had 21–28 small, pointed, slightly curved teeth in each jaw.

NATURAL HISTORY: Pacific white-sided dolphins can form huge shoals of a thousand or more, though groups of several hundred or fewer are far more common. Herds usually contain individuals of all age groups and both sexes. Throughout their range, Pacific white-sided dolphins are gregarious, and appear to seek out the company of many other marine mammals. Their most common association is with the northern right-whale dolphin, with which they share a nearly identical range.

The calving season has been reported to be summer, but in recent studies calves have been found primarily in early fall.

Pacific white-sided dolphins race alongside a Japanese fisheries research vessel. (Northern central Pacific, near Attu Island, July 1978: David Ambrose, U.S. National Marine Fisheries Service.)

Seasonal movements are not well understood in most areas. In some areas, such as off Monterey, California, and off southern California and northwestern Baja California, there appear to be resident pods that are augmented during fall through spring by influxes of animals from other areas, thought to be farther north and offshore.

White-sided dolphins are extremely animated, often leaping clear of the water and engaging in acrobatics. Running herds moving at a moderate pace, apparently undisturbed by the presence of observers, have been seen to churn the sea surface with frequent leaps and gambols.

The species feeds on a wide variety of fish and squid, primarily (judging by radiotelemetry data) at night. They often can be seen at dawn and dusk feeding with gulls on small surfacing balls of unidentified bait fish. So far squid, anchovies, and hake have been most prevalent among the stomach remains examined.

DISTRIBUTION AND CURRENT STATUS: Pacific white-sided dolphins have a primarily temperate distribution, remaining north of the tropics and south of the colder waters influenced by arctic currents. They have been reported from the Kamchatka Peninsula, Amchitka Island, and Kodiak Island south into the Sea of Japan and along the entire Pacific coast of Japan (on the west) and south to the tip of Baja California, with stragglers into the Gulf of California (on the east). Most sightings of the species have been in waters between the seaward edge of the continental slope and the 100-fathom curve, though closer approaches to shore have been noted where deep water is nearby.

Nowhere are Pacific white-sided dolphins the object of an important fishery. Some are taken for food each year in coastal fisheries off Japan, but they are difficult to drive; so they do not figure in any significant way in Japanese drive fisheries. During the last 25 years, small numbers have been captured live, mostly off southern California, for display in zoos and aquariums, and smaller numbers have been taken in Japanese waters for aquariums. Recently some have been killed accidentally in the North Pacific drift-net and gill-net fishery for salmon.

Two stocks have been proposed: northwestern Pacific and northeastern Pacific, apparently separated by an area of low density along the south side of the central Aleutian Islands. Morphological differences appear to exist between animals in the northeastern Pacific north of southern California and those "resident" off Baja California. Size of the population(s) is not known.

CAN BE CONFUSED WITH: Within their exclusively temperate North Pacific range, Pacific white-sided dolphins are most often mistaken for common dolphins, but particularly fast-moving animals may resemble Dall's porpoises. Common dolphins have a smaller, more gently back-curved dorsal fin, a moderately long beak, which is usually white-tipped, and a complicated color pattern that includes an hourglass configuration to the pigmentation pattern on the sides, a dark band from the middle of the lower lip to the flipper, and usually subtle markings on the face and flanks. They are rarely found north of central California. Dall's porpoises have a relatively low, triangular, and bicolored dorsal fin, a head that is very small in proportion to the chunky body, and stark black-and-white coloration in clearly demarcated zones. Though both Pacific white-sided dolphins and Dall's porpoises may produce a rooster tail of spray as they slice through the water, the latter's dorsal fin should obviate confusion. Furthermore, Dall's porpoises do not leap clear of the water.

Hourglass Dolphin

Lagenorhynchus cruciger
(Quoy and Gaimard, 1824)
DERIVATION: from the Latin
crucis for "cross," and *gero* for
"bear, carry."

ZONES 3 TO 6

DISTINCTIVE FEATURES: Sharply demarcated black-and-white coloration, with dorsal and ventrolateral dark zones meeting at midflank, separating white sides into anterior and posterior zones; found only in high latitudes of southern hemisphere.

DESCRIPTION: The largest measured specimens were a 163-cm male and a 183-cm female, but too few have been examined to allow any conclusion about size ranges.

This is a robust dolphin, with a short, thick, well-defined beak. The dorsal fin, as is typical of the genus, is prominent, wide at the base, and strongly recurved. The flippers are modestly long, tapered, and backswept. There is a thick keel on the caudal peduncle, above and below.

The hourglass dolphin has a striking black-and-white color pattern. The snout and dorsal surface are dark, including the uniformly black dorsal fin and tail. The sides of the face, the throat, and the belly are white. The sides and flanks are bicolored, with a black eye band extending lateroposteriorly to encompass the flipper, and widening on the side just ahead of the dorsal fin, at

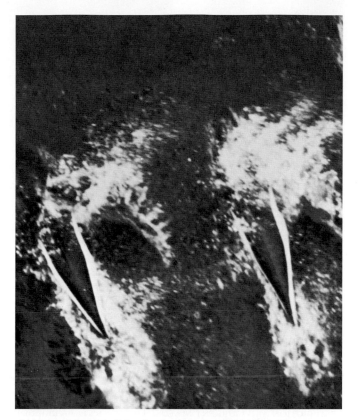

A rare dorsal view of the lightening-fast hourglass dolphin, bow-riding. (Northeast of Scott Island, Antarctica, 66° 36'S, 177° 51'E, January 23, 1981: Gerry Joyce.)

which point it abuts the dark dorsal field. Posterior to the fin, the dark areas above and below attenuate markedly, allowing a white patch on the side to taper to an end near the insertion of the flukes. The lower of the dark regions is broken by irregular white intrusions above the anus. There are about 28 teeth in each row.

NATURAL HISTORY: Groups of six to forty dolphins have been sighted. Nothing is known about their reproductive biology or seasonal movements. Hourglass dolphins have been seen swimming with fin whales in the Antarctic, and in the close company of pilot whales, southern bottlenose whales, and Arnoux's beaked whales. They are known to ride bow waves during rough weather, occasionally spinning around their longitudinal axis and performing short vertical jumps through the crests of large waves.

At sea, a group of these dolphins resembles a flock of penguins

The most southerly of dolphins, hourglass dolphins are a pleasant surprise to seafarers sailing in high latitudes of the Southern Ocean. (South of the Antarctic Convergence, about 50°S, 30°W, January 1970: Peter Beamish.)

in that, when seen at a distance, both undulate in a smooth wave motion. Hourglass dolphins are avid bow riders, and can be seen to make long, high leaps as they rush for a ship's bow. When avoiding vessels, they remain near the surface, even when submerged, and then breathe amid a confused spray of their own creation. Food habits and mortality factors are unstudied.

DISTRIBUTION AND CURRENT STATUS: This species is circumpolar in the Antarctic and subantarctic. It can be expected north or south of the Antarctic Convergence, and in cool currents associated with the West Wind Drift. It is the only dolphin likely to be encountered in the Antarctic that has a dorsal fin (the finless southern right-whale dolphin may be seen there, particularly south of Cape Horn). Although it has been sighted off the Chilean coast as far north as 33°40′ S, the vast majority of records are from between 45° S and 65° S. Not only does it show a marked preference for colder waters, but it is truly pelagic and rarely seen near land.

Although widely distributed, nothing is known about this dolphin's abundance. It has never been exploited, intentionally or accidentally, to the best of our knowledge.

CAN BE CONFUSED WITH: There is a good possibility of confusing this species with the dusky dolphin and Peale's dolphin (off South America and the Falklands only). The fact that it is a coldwater, offshore animal can help identify it, but careful attention must be paid to coloration differences. Photographs will often be essential to good field identification of this and its allied species.

Peale's Dolphin

Lagenorhynchus australis
(Peale, 1848)
DERIVATION: from the Latin
australis for "southern."

DISTINCTIVE FEATURES: Similar to *L. obscurus,* but coloration muted; black face and throat; conspicuous dark, diagonal flank patch, above which is a white patch narrowing anteriorly; small white patch above and behind flipper; found only in coastal waters of southern South America and in the vicinity of the Falkland Islands.

DESCRIPTION: The largest measured specimen was 216 cm long. The color pattern distinguishes this dolphin from the other two southern-hemisphere "lags." The dorsal surface, dorsal fin, flippers, and flukes are dark. A continuous black patch covers the lower lip, chin, and throat, narrowing and ending below the eyes. The sides are light gray, except for a broad, diagonal patch beginning with the dark dorsal field anterior to the dorsal fin and ending above the anus, where it blends with the dark underside of the tail stock. There are, then, two light gray areas on each side, much as in the white-beaked dolphin of the northern hemisphere. The white abdominal field is sharply demarcated from the gray side by a low, irregular line from the anus to the eye. There is a small white patch immediately behind the axilla. Few specimens have been examined and reported, but there are about 30 teeth in each row.

NATURAL HISTORY: We know of only two useful observations on the species' natural history. Groups seen at the Falkland Islands and at Ushuaia, Argentina, were swimming and diving among beds of giant kelp. In the latter sighting the dolphins played near the vessel while it was in the kelp, but abandoned it each time it moved into open water. One individual from the group seen at the Falkland Islands was collected, and it had remains of octopus in its stomach. These dolphins are known to leap clear of the water, and occasionally to bow-ride.

DISTRIBUTION AND CURRENT STATUS: This species is confined to coastal waters of southern South America. There are no records of it farther north than Valparaiso on the Pacific coast and about Golfo San Matias on the Atlantic coast, and none offshore, except at the Falkland Islands, where it is common, and in the waters south of Cape Horn to about 57° S. Magellan Strait and the southern coast of Tierra del Fuego are other areas in its range. It is common in the Beagle Channel as well. Off southern Chile Peale's dolphins were formerly harpooned for crab bait. It is not known whether this practice, now prohibited, continues. Some of these dolphins have reportedly also been taken incidentally in gill nets for crabs in Chilean waters and for fish in Argentinian waters.

CAN BE CONFUSED WITH: Peale's dolphin may be confused throughout its range with the dusky dolphin and possibly the hourglass dolphin. Its black face and throat, and the two diagonal flank patches, one white, one black, may be the best keys for distinguishing it from them. Also, the white abdominal field extends above the flipper by way of the axilla in Peale's dolphin, but in the other two species the dark flippers are immediately adjacent to the black or dusky pigmentation of the flanks. These features should make the species relatively easy to identify when seen at close range.

Fraser's Dolphin

Lagenodelphis hosei
Fraser, 1956
DERIVATION: from the Greek
lagenos for "bottle," and *delphis*
for "dolphin," referring to a
combination of characteristics
of *Lagenorhynchus* and
Delphinus; C. Hose was the

Sarawak resident who collected
the type specimen in 1895.

ZONES 1 TO 5

DISTINCTIVE FEATURES: Short beak; relatively small flippers, flukes, and dorsal fin (subtriangular); wide, prominent eye-to-anus black band; stocky build; tropical distribution.

DESCRIPTION: An adult female measured 236 cm and weighed 164 kg; an adult male, 264 cm and 209 kg. Length at birth is about 1 m.

Morphologically, this species has features of both *Lagenorhynchus* and *Delphinus;* so it was given the generic name *Lagenodelphis*. It has a robust body with a short beak (though the mouthline is long), and rather small dorsal fin and flippers relative to body size. The flippers taper to a point; the flukes have a concave trailing edge; and the slender, centrally situated dorsal fin is pointed at the peak and weakly falcate, although sometimes the rear margin is vertical or even convex.

The body is bluish gray on the back and white on the belly, with a longitudinal striping pattern on the sides. A cream-white band begins above and forward of the eye, and extends along the flank, becoming indistinct above the anus. There are complex markings on the face. A parallel black band begins around the eye (below the white band) and extends to the anus. This is the most prominent color characteristic, and its width and intensity apparently increase with age. A second cream-white band below and parallel to the dark stripe separates the darker gray of the sides from the white ventral field. A light gray stripe (or pair of stripes) connects the gape of the mouth with the anterior insertion of the flipper. The appendages, including the upper lip as well as the flippers, fin, and flukes, are dark.

Some 34 to 44 pairs of slender, pointed teeth line the upper and lower jaws.

NATURAL HISTORY: Groups of a hundred to a thousand have been seen; the species appears to be highly gregarious. Occasionally these dolphins will be found mixed in herds of spotted dolphins, and they have been observed in the company of false killer whales, sperm whales, striped dolphins, and spinner dolphins as well. Nothing is known of their reproductive biology or seasonal movements. Their swimming style is aggressive; they often create a spray as they charge to the surface to breathe. They are known to leap clear of the water, but are less acrobatic than many other pelagic dolphins. In some parts of their range, they are regarded as shy and difficult to approach, but not off South Africa, where they commonly ride the bow waves of vessels. Several have been taken alive in the Philippines for display, but so far their survival record has not been good.

Their diet consists of squid, crustaceans, and deep-sea fish, which are captured at night when they rise toward the surface.

DISTRIBUTION AND CURRENT STATUS: This tropical pelagic species was first described in 1956 from the remains of a beach-washed spec-

Fraser's dolphins, discovered scarcely 25 years ago, are tropical creatures of the high seas. (Eastern equatorial Pacific, 00° 22′ N, 95° 45′ W, January 16, 1979: Robert L. Pitman.)

imen found at Sarawak (Malaysia) in the South China Sea. Since then, it has been found stranded or observed at sea in many areas across the tropical Pacific, off eastern Australia, in the Indian Ocean off South Africa, and off Japan and Taiwan. In the tropical eastern Pacific, it is most often found in equatorial waters. Recently specimens have been reported from the Lesser Antillean island of St. Vincent in the tropical Atlantic. The species may be considered an inhabitant of low latitudes in all three major ocean basins.

Fraser's dolphin has been exploited to an unknown extent in the eastern tropical Pacific purse-seine fishery for yellowfin tuna. When they are taken, these dolphins are usually mixed in incidental catches of spotted dolphins. Individuals are captured at least occasionally by the pilot-whale hunters of St. Vincent.

CAN BE CONFUSED WITH: There are several species with which this one may be confused. The striped dolphin is probably the most difficult to distinguish at sea. However, its black lateral stripes are narrower and more sharply demarcated, and its forward-pointing tongue of dark pigmentation at midflank is conspicuous. Also, the striped dolphin is more slender, and has a much longer beak, longer flippers, and a taller, wider-based dorsal fin. The common dolphin's long beak and hourglass pigmentation pattern on the flanks should be adequate to distinguish it. In the North Pacific, the Pacific white-sided dolphin may cause confusion. However, it has a less distinguishable beak, longer flippers, and a large, strongly recurved dorsal fin.

Common Dolphin

Delphinus delphis
Linnaeus, 1758
DERIVATION: from the Latin
delphinus for "dolphin,
porpoise"; from the Greek
delphis for "dolphin."

ZONES 1 TO 5

DISTINCTIVE FEATURES: Conspicuous white thoracic patch;
V-shaped black or dark-gray saddle with downward-oriented apex
on sides directly below dorsal fin; light gray of flank sweeping over
dorsal aspect of tail stock; hourglass effect on side, with tan or yellowish tan region making up posterior half of hourglass; absent
from high latitudes, but otherwise cosmopolitan.

DESCRIPTION: Body length ranges to about 2.5 m, but most adults
are 2.3 m or less, with males slightly larger than females. They
seldom weigh more than 75 kg. Newborn are about 80 cm long.
   The common dolphin is, as its scientific name implies, the ar-
chetypal dolphin. In appearance and behavior it is everything a dolphin is supposed to be. The body is slender, streamlined, elegant.
The long, well-defined beak is generally black, but often tipped
with white. The lips are black. The prominent dorsal fin, situated at
midback and triangular or falcate in shape, ranges in color from all
black to mostly white with a dark border; there is often a grayish
area in the middle of an otherwise dark fin. The flippers, which
taper to a pointed tip, and flukes, which are concave with a distinct

median notch on the rear margin, are dark gray or black. A dark stripe connects the dark eye-patch with the corner of the mouth, and another connects the lower jaw with the flipper.

Although the head, beak, and dorsal fin are generally similar to those of several other dolphins (mainly species of the genus *Stenella*), the color pattern on the common dolphin's body is very distinctive. The back is black or brownish black; the chest and belly white or cream. The dark cape dips most deeply onto the sides just below the dorsal fin, where it meets the upward intrusion of the ventral whiteness. The V-shaped saddle formed thereby often serves as the most useful field mark during sightings at sea. The joining on the sides of the contrasting dorsal and ventral coloration creates a crisscross or hourglass effect, with two adjacent zones of white, gray, grayish green, or yellowish tan on each side of the animal. The posterior zone on either side sweeps as a broad gray swath onto the dorsal aspect of the tail stock.

It should be noted that many pigmentation features, particularly the intensity of the striping, differ by region, as do such characteristics as length of beak and body size. There are 40–55 (perhaps as many as 58) small, sharply pointed teeth in each row.

NATURAL HISTORY: Common dolphins are among the most gregarious of cetaceans. They are found regularly in herds numbering many hundreds, sometimes more than a thousand. It is not unusual in the eastern North Pacific to see them mixed with Pacific white-sided dolphins on productive feeding banks, but the two species generally do not intermix while traveling. They ride the bow of both powered vessels and large whales; we have seen them frolicking in the bow waves of fin whales in the western North Atlantic, and of gray, fin, humpback, and blue whales in the North Pacific. When met under the right circumstances, they will stay with a vessel for long periods.

The common dolphin's life history has been studied most extensively in the eastern North Pacific. Here calving appears to peak in spring and fall. Gestation lasts for ten to eleven months. The average interbirth interval is more than one year. There is some evidence that lactating and near-term pregnant females remain segregated from the rest of the population.

In most of their range, common dolphins are pelagic, offshore creatures that are most likely to be encountered along or seaward of the 100-fathom contour. Off southern California they are associated with conspicuous features of bottom relief, such as sea mounts and escarpments, preying at night on organisms associated with the deep-scattering layer (DSL). They can dive to depths of at least 280 m, and for as long as eight minutes. The daily activity cycle of a large herd generally begins in late afternoon, when small feeding groups scatter to await the ascension of the DSL. At dawn, or later on overcast days, as the light-sensitive scattering layer returns to

When common dolphins are predisposed to play, only the fleetest of passing boats can evade them. (Northwest of Point Conception, California, 35°55' N, 123°49' W, June 29, 1980: Robert L. Pitman.)

A sprinting group of sleek common dolphins beside a research vessel. (Off Oman, January 15, 1982: Hal Whitehead.)

depths out of reach, these groups coalesce for periods of rest and social interaction, in preparation for the resumption of feeding late in the day.

When they strand, common dolphins are usually alone; there certainly is no evidence that they strand in large groups like certain other gregarious dolphins and whales do. There is an interesting account of a small herd of several dozens of animals moving far up the Hudson River in New York State during the month of October. At least two eventually washed ashore, one 135 km, the other 270 km, from the sea.

Well-defined migrations are not known, and common dolphins are present the year round in some parts of their range. However, clear seasonal shifts in distribution are observed off southern California, for instance, where peaks of abundance occur in June, September through October, and January. Water temperature appears to be significant, with 10° to 28° C apparently preferred in the eastern North Pacific.

In addition to their bow-riding antics, common dolphins are aerial acrobats. Melville probably had them in mind when he wrote "they always swim in hilarious shoals which upon the sea keep tossing themselves to heaven like caps in a Fourth of July crowd." In spite of their frolicsome nature in the wild, common dolphins do not generally adapt well to captivity and are rarely exhibited alive.

Their diet consists primarily of fish and squid. Off southern California, anchovies and squid are especially important in winter; deep-sea smelt and lanternfish predominate in spring and summer. Like other cetaceans, common dolphins sometimes take advantage of human fishing activities by feeding on disabled fish escaped or discarded from nets.

DISTRIBUTION AND CURRENT STATUS: The common dolphin is very widely distributed, occurring in all oceans to the limits of tropical and warm temperate waters. There are several distinctive forms that probably deserve racial or subspecific status; some scientists recognize more than one species.

Its pelagic range is so farflung that there is little point in noting details about it. There are various recognized local forms in the eastern North Pacific, the Mediterranean Sea, the Black Sea, along the European and African Atlantic coasts, in the Indian Ocean, and off Japan. Such recognition, however, probably says as much about the distribution of investigators as about actual dolphin distribution.

In the South Pacific there are records for the South American coast from the equator to 40° S off Chile.

The species has been hardest hit in the Black Sea, where a major Soviet direct fishery took as many as 120,000 per year at one time, and in the eastern tropical Pacific, where a multinational tuna purse-seine fishery accounted for approximately 8,000 in one year in the early 1970s. The status of common dolphins in the Black Sea is a source of concern. They certainly were depleted there by the mid-1960s, after which the Soviet government closed the fishery, but a Turkish dolphin fishery in the Black Sea has persisted. The tuna fishery's impact on common dolphins is thought to be less severe than it is on other dolphin species (e.g., *Stenella* spp.), particularly since common dolphins are notoriously difficult to catch in much of the fishing zone. Throughout the world, a few common dolphins are killed directly in harpoon or drive fisheries, and inadvertently in fixed or active fishing gear. Also, many have expressed concern about reports of "sportsmen" shooting at common dolphins for fun in the Mediterranean.

CAN BE CONFUSED WITH: Of the many species of oceanic dolphins whose ranges overlap portions of the common dolphin's range, striped dolphins are the most troublesome. The saddle below the dorsal fin of the common dolphin and the complex pattern of dark stripes (eye to anus and eye to flipper) of the larger striped dolphin are the primary clues to separating them in the field.

Care should be taken in tropical portions of their range not to confuse common dolphins and short-snouted spinners. When present and seen clearly, the hourglass pattern on the common dolphin sets it apart from all other species.

Bottlenose Dolphin

Tursiops truncatus
(Montagu, 1821)
DERIVATION: from the Latin
tursio for "an animal like the
dolphin" (from Pliny), and the
Greek *ops* for "face"; from the
Latin *truncare* for "cut off."

ZONES 1 TO 5

DISTINCTIVE FEATURES: Beak usually short and stubby but sometimes moderately long; head and body robust; color pattern a subdued blend of brown to charcoal; generally nondescript cape,
lighter sides, still lighter belly; dorsal fin moderately tall and falcate;
absent only from polar regions.

DESCRIPTION: In all areas where the systematics of bottlenose dolphins have been studied, there appear to be two ecotypes, a coastal
form and an offshore form, including residents of coastal and
oceanic islands. Differences between them are currently under
study by W. A. Walker and W. F. Perrin, but no consistently reliable differences that would allow the forms to be distinguished in
the field are known so far. The many nominal species, including
perhaps *T. aduncus*, may simply reflect differences within these two
forms, whose occurrence is well-documented. A taxonomic re-
vision of the genus is needed – but that is not within the scope of
this book.

Maximum size reliably reported is 3.9 m and 275 kg, though
larger size estimates have been published for some southern-hemi-

Mighty mid-ocean swells serve as a playground for bottlenose dolphins, including this mother and her large calf. (Off Maui, Hawaii, April 1980: Bernd Würsig.)

sphere populations. Males are larger than females. Length at birth is about 0.9 to 1.2 m.

The body shape differs within and between regions, and is difficult to characterize simply. In most areas the majority of animals have a short, stubby beak and a robust, almost chunky, head and trunk, slimming rapidly behind the dorsal fin. But much slimmer bottlenose dolphins, sometimes with longer, slimmer snouts, are often found in the same group. The snout is always clearly demarcated from the forehead by a sharp crease or apex.

The flippers are of moderate length, pointed on the tip. The dorsal fin, which is moderately tall and falcate, is located about midback, though the fact that the forepart of the body is so much more robust than the aft makes the fin appear to be slightly behind the midpoint. The flukes are sinuously curved on the rear margin and deeply notched.

Coloration can range through many shades of charcoal, lighter gray, and brown. The body is marked by a nondescript cape, beginning at the apex of the melon, gradually broadening from the blowhole to the dorsal fin, then narrowing to a thin line behind the dorsal fin. The sides are lighter than the cape, and the belly lighter still, from light gray to pink. There is no clear line of demarcation between side and belly coloration. Spots are present on some animals. Though present, subtle markings on the face and throat, and an eye-to-flipper stripe, are barely visible on most individuals. Color variants, including all black, all white, and cinnamon-colored bottlenose dolphins, occur.

To date, specimens have been reported with 18 to 26 teeth in each row. Teeth are often well worn in older specimens, particularly in coastal forms, probably reflecting differences in diet and feeding habits between the two forms.

Natural History: Bottlenose dolphins usually form groups of fewer than ten (coastal) or 25 (offshore) individuals, though herds of several hundred have been reported from some offshore regions. Males reach sexual maturity at ten to twelve years of age, females five to twelve. Once reproductively active, females bear a single calf every second or third year. Gestation is about twelve months.

Calves are often nursed for a year or more. They are closely tended by adults during at least the first half-year. "Babysitting" has been observed, in which nearby adults remain with a calf as its mother forages for food. As they age, young dolphins become more independent, tasting "scraps" of food in the wake of feeding adults and gradually learning to hunt on their own. Echolocation and other feeding-related behavior probably is learned during this prolonged period of weaning.

Everywhere they have been studied, coastal bottlenose dolphins appear catholic in their feeding habits, taking a wide variety of fish and invertebrates. In many areas they have adapted their feeding strategies to take advantage of human activities, eating netted fish, fish discarded as trash by fishermen or stirred up by nets and propeller washes, and fish attracted to idle vessels and fixed platforms. In shallow areas, and near the surface in others, they often invert and feed upside-down, presumably to aid in echolocation by reducing noise from surface echoes. Various kinds of cooperative hunting have been reported. Offshore bottlenose dol-

Cooperative fishing: bottlenose dolphin (Tursiops) *and a Mauritanian villager. (Cap Timiris, Mauritania: courtesy of René Guy Busnel.)*

phins show a decided preference for squid, which is no doubt why they often associate with herds of the squid-eating pilot whales.

Bottlenose dolphins are capable and willing bow riders, often charging to a moving vessel from great distances to frolic in its wake. They often also ride ground swells and pressure waves of big whales. In most areas where they occur near shore, they frequently can be found body surfing as well.

Bottlenose dolphins are the most common cetacean in captivity, and have been the mainstay of displays at zoos, aquariums, and marine parks around the world since the first were successfully held captive in 1914. Since they seem to be adaptable generalists, it is not very surprising that bottlenose dolphins have mated and produced hybrid offspring with a rough-toothed dolphin, a pilot whale, a false killer whale, and Risso's dolphins, the last both in captivity and possibly free-ranging. Stories of dolphins befriending people have usually been about this species, since its coastal haunts and curiosity often bring it into close contact with humans. Some bottlenose dolphins have become "residents" of particular areas, where they sometimes allow themselves to be touched.

DISTRIBUTION AND CURRENT STATUS: Bottlenose dolphins are cosmopolitan, avoiding only the very high latitudes. In the Pacific they range from northern Japan and southern California to Australia and Chile; in the Atlantic from Nova Scotia and Norway to Patagonia and the tip of South Africa, being fairly common in the Mediterranean Sea; in the Indian Ocean from Australia to South Africa. The coastal form is usually found shoreward of the 10-fathom curve and frequently enters harbors, bays, lagoons, estuaries, and river mouths, often ascending rivers for several scores of miles. There is mounting evidence that these coastal animals have limited home ranges along overlapping segments of coast. The offshore form appears less restricted in range and movements, being present in many productive areas, particularly in the tropics.

Throughout their ranges, both forms seem strongly attracted by human activities and in some areas are shot as a nuisance to fishermen. Direct exploitation is limited. In the Black Sea, a commercial fishery for oil and fishmeal depleted the dolphin population severely by the mid-1960s, after which the Soviet Union terminated its harvest. Hunting by Turkey continues, however, and unknown numbers are taken annually. A few are taken for food in Sri Lanka, West Africa, Venezuela, the West Indies, and other parts of the world. The United States has had a controlled live-capture fishery for the species in operation since 1938. Most animals have been taken off eastern Florida, in the Gulf of Mexico, and off southern California. Scattered live specimens have also been captured in recent years off Hawaii, South Africa, Japan, Mexico, the Philippines, and the Bahamas, and in the Mediterranean. A few individuals are killed incidentaly by pelagic seining and beach trammel-net and gill-net fishing.

CAN BE CONFUSED WITH: In the more pelagic portions of their range, bottlenose dolphins are most likely to be mistaken for rough-toothed dolphins, young Risso's dolphins, and spotted dolphins. Rough-toothed dolphins have a long, cone-shaped head with no apex to the melon and an extremely thin cape. Risso's dolphins have a blunted, beakless head, and a taller, more falcate dorsal fin; they generally lack a visible cape altogether, and lighten with age from slate gray toward white, beginning on the head. As the name implies, spotted dolphins have spots, though their extent and intensity differ with age and region. Even if spots are absent, spotted dolphins examined at close range can be seen to have a more complex color pattern and a longer, slimmer beak. Oceanic bottlenose dolphins might also be mistaken for the more lethargic dwarf sperm whale, though under most circumstances the two should be easily distinguished. Dwarf sperm whales are usually encountered lying at the surface, like so many logs afloat, and as a rule settle quietly out of sight, not rolling as most dolphins do. At close range the bluff, beakless forehead is a further key.

In northern South America, care should be taken in coastal areas not to summarily report a sighting of a small gray dolphin in riverine, estuarine, or nearshore habitats as a bottlenose dolphin. The smaller (to 1.7 m), similarly shaped tucuxi is found here, and close approach may be necessary to distinguish the two species with certainty. In the Indian Ocean and on the west and south coasts of Africa, bottlenose dolphins might easily be mistaken for young hump-backed dolphins, which apparently lack the pronounced hump of adults. East of Indonesia, even adult hump-backed dolphins lack the hump; so here confusion is likely. The pale body color and triangular dorsal fin of adult hump-backed dolphins east of Indonesia should make them distinguishable from bottlenose dolphins.

(See the section on striped dolphins for a discussion of superficial similarities between bottlenose, spotted, and striped dolphins.)

Risso's Dolphin

Grampus griseus
(G. Cuvier, 1812)
DERIVATION: from the Latin
grandis for "great" and *piscis* for
"fish," giving rise to the New
Latin *grampus* for "a kind of
whale"; from the Middle Latin
griseus for "gray."

ZONES 1 TO 5

DISTINCTIVE FEATURES: White or light-gray coloration of adults, usually interrupted only by dark dorsal fin, flippers, and flukes; tall, falcate dorsal fin; extensive scarring on adults; lack of beak; squarish melon bisected by deep crease; primarily found in deep tropical and warm temperate waters worldwide.

DESCRIPTION: Adult Risso's dolphins are 3.6 to 4 m long, and they are not known to be sexually dimorphic. Newborn are about 1.5 m long.

The body is robust anterior to the dorsal fin, and the blunt snout lacks a noticeable beak. A peculiar, deep, V-shaped furrow is present in the area of the forehead, but usually can be seen only when an animal is inspected at close range.

The appendages are very prominent: long, pointed flippers, and a tall, slender, falcate dorsal fin, placed approximately at midback.

This species' most conspicuous external features are the coloration and markings of adults. Newborn are an even light gray, and they change to an even chocolate brown early in life. By adulthood, however, much of the pigment has been lost, and parts of the body are white or light gray except for a narrow cape, which is sometimes visible. The dorsal fin and an area of the back adjacent to its base remain dark, as do the flippers and flukes. The head becomes almost completely white, with a dark area around each eye. Scarring is extensive; in fact, adults usually have a battered appearance, looking as though a bucket of white paint has been spilled on them. Appreciation of their whiteness depends to some extent on lighting conditions. There is a large anchor-shaped light gray patch on the dark ventrum (similar to that found on pilot whales).

The dentition of this species is peculiar; generally there are no maxillary teeth and no more than seven pairs of peglike mandibular teeth. These teeth are often badly worn in older individuals, and some can be missing entirely.

NATURAL HISTORY: Risso's dolphins are occasionally seen as solitary individuals and pairs, but they usually are more gregarious, living in herds of 25 to several hundred. Sometimes they swim in "echelon" formation, lined up abreast at evenly spaced intervals, a tactic that is probably effective in the search for prey.

It is not unusual to find these deepwater, pelagic dolphins in close company with other oceanic cetaceans, particularly pilot whales. They show a marked preference not only for water greater than 100 fathoms deep, but also for warm temperate to tropical conditions, though some do wander into cooler latitudes in summer.

There has been no major fishery for Risso's dolphins; so their life history has not been studied in detail. Males are believed to attain sexual maturity at lengths of 3 m or more. A few have been caught alive and maintained for long periods in captivity in Japan, and aquariums in the United States have had some success keeping live-stranded specimens alive for several years. One of those captive in Japan mated with a bottlenose dolphin and produced a hybrid calf. Suspected hybrids, with characteristics of Risso's dolphins and bottlenose dolphins, were observed on the Irish coast in the 1950s.

Their principal prey is squid, and they probably eat fish at least occasionally.

Risso's dolphins are active but not as acrobatic as many other dolphins. They flipper- and fluke-slap, spyhop, and make breaching body slams during periods of intense activity, but at other times they simply roll at the surface. It is usually almost impossible to entice Risso's dolphins to bow-ride for more than a few seconds. However, a famous individual known as Pelorus Jack "escorted" boats into a New Zealand harbor for more than twenty years, and apparently rode bow waves regularly.

DISTRIBUTION AND CURRENT STATUS: Risso's dolphin is abundant and cosmopolitan in tropical and warm temperate latitudes. The

A trio of Risso's dolphins bursts through the surface of the tropic sea. (Eastern tropical Pacific, June 1980: Robert L. Pitman.)

extremes of its range are generally taken to be from Newfoundland and Sweden south to the Mediterranean and the Lesser Antilles in the North Atlantic; as far south as Argentina and South Africa in the South Atlantic; from offshore southeastern Alaska and the Kuriles south to central Chile in the Pacific, and to New Zealand and Australia in the Indo-Pacific. Off western North America, where the species' distribution has been studied in some detail, there are two apparent gaps where Risso's dolphins are virtually absent, centering at about 20° N and 42° N.

Risso's dolphins are hunted only incidentally in small cetacean fisheries around the world. There are catch records for Newfoundland, the Lesser Antilles, Japan, the Solomon Islands, and Indonesia. They are occasionally killed inadvertently in fishing gear in many areas, though the mortality is not well-documented and regarded as negligible.

CAN BE CONFUSED WITH: Risso's dolphins have very distinctive features and can be identified easily at close range. However, at a distance they can be confused with bottlenose dolphins, false killer whales, and females or immature male killer whales. The rounded head should distinguish them from bottlenose dolphins, as should the lightness of adult coloration. False killer whales are not so extensively scarred and have uniformly dark pigmentation. Killer whales are marked by well-defined areas of white on a shiny black background, and have a more tapered head than Risso's dolphins.

Because of their partial whiteness and lack of a prominent beak, Risso's dolphins are often misidentified by untrained observers as white whales. Their tall and prominent dorsal fin, however, should rule out this kind of mistake. Cuvier's beaked whale is another pale species that could cause problems, particularly since its tropical to warm temperate range is much like that of Risso's dolphin. Its smaller and more posteriorly positioned dorsal fin, larger adult body size, and tapered head should help differentiate it.

Spotted Dolphin

Stenella spp.,
including *Stenella attenuata*
(Gray, 1846); *Stenella plagiodon*
(Cope, 1866); *Stenella frontalis*
(G. Cuvier, 1829); and *Stenella
dubia* (G. Cuvier, 1812).
DERIVATION: *Stenella* from the
Greek *stenos* for "narrow,"

apparently referring to the long,
narrow beak.

SOUTHERLY PORTIONS OF
ZONES 1 AND 2, NORTHERLY
PORTIONS OF ZONES 3 TO 5

STENELLA ATTENUATA

STENELLA PLAGIODON

DISTINCTIVE FEATURES: Body usually spotted, differing by region and age; spotting generally decreases with distance from continental shores of North America, but within populations increases with age; beak long and slim; distribution primarily tropical but includes some warm temperate waters in Atlantic; distribution primarily oceanic.

DESCRIPTION: Whatever the names that may eventually be used for them, there appear to be two species of spotted dolphin in the Atlantic and Pacific oceans. The coastal forms in each species are larger and more robust, and have a greater fluke span. Offshore forms are more elongated, as is typical of other oceanic dolphins of the genera *Delphinus* and *Stenella*. Spotting develops with age in all populations, albeit variably, and is most extensive in large adults. In general the spotting is heaviest in the coastal forms off both coasts of North America. As one moves eastward in the Atlantic and westward in the Pacific, the degree of spotting in adults decreases. Hawaiian spotted dolphins are usually unspotted; Japanese and eastern Indian Ocean animals have only moderate spotting. At least in the Pacific, these morphological differences warrant racial but not species status.

Male Pacific spotted dolphins reach approximately 2.3 m (offshore), 2.5 m (coastal); females 2.2 m (offshore), 2.3 m (coastal). Atlantic animals (of both sexes) reach 2.2 m. Length at birth is 80 to 90 cm.

Body form differs slightly with region, coastal animals tending to be larger and more robust. The species has a moderately long beak, a tall, falcate dorsal fin, and a slightly noticeable keel, except perhaps in large adult males. Neither the flippers nor the flukes bear any unusual contours, resembling those of many other oceanic dolphins.

Coloration in the Indo-Pacific forms is complex. There is a dis-

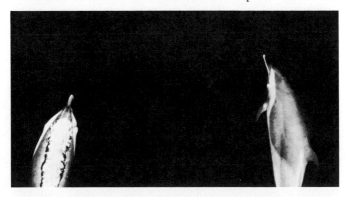

Spotted dolphins photographed by strobe light as they ride the bow near midnight. (Southeastern Arabian Sea, February 10, 1982: Abigail Alling— WWF.)

tinct cape from dark gray to brownish gray, narrow on the forehead, dipping low below the dorsal fin, then swinging upward, to end about the middle of the tail stock. (*S. plagiodon* varies from this pattern, in that the cape extends farther aft on the tail stock.) The dorsal fin is the same color as this cape. There is a complex pattern on the face also, consisting of an eye patch, an eye band, a border around the gape of the mouth, and a flipper band. These are dark but sometimes indistinct, particularly in larger animals. The lips and the tip of the snout are often white. Amount of spotting increases with age, beginning on the head and ventrum, then spreading rapidly over the entire body. In both the Atlantic and Pacific, coastal forms are most heavily spotted, with little dorsoventral contrast. In the Pacific, the spots of Hawaiian spotters are so faint they seem absent, so that these dolphins resemble neonates of the offshore race. Some spotted dolphins in the North Atlantic have a shoulder blaze similar to that found on striped dolphins.

The spotted dolphins are among the most difficult cetaceans to describe. There are many geographical forms, and probably only two valid species, but these are variously referred to by the four names listed; the first is assigned to spotted dolphins in the Pacific, Indian, and Atlantic Oceans, the other three to animals in the Atlantic only. The complicated problem of unraveling the spotted dolphins' taxonomy and describing their biology has been undertaken jointly by W. F. Perrin, E. D. Mitchell, and P. J. H. van Bree, and it is primarily their work that is summarized here.

NATURAL HISTORY: These comments are based principally on eastern Pacific data, and may not be completely accurate for spotted dolphins from other areas.

Spotted dolphins are sometimes found in herds of a thousand or more in offshore areas, though groups of a few hundred are more common, and herds now appear smaller than previously in the eastern tropical Pacific. Coastal groups are generally much smaller, sometimes containing a few hundred individuals, more frequently fifty or fewer. Herds generally contain individuals of both sexes and all age classes.

In all areas, but particularly offshore, spotted dolphins are gregarious, frequently schooling with other species, especially spinner dolphins in the tropical Pacific. It is below such mixed aggregations of spotters and spinners that schools of yellowfin tuna and (less frequently) skipjack tuna congregate for reasons still not completely understood.

The average male reaches sexual maturity by 200 to 207 cm, at an estimated six to eleven years; for females it is 187 to 195 cm, at about 4.5 to 8.5 years. Calves are born throughout the year, with multiple peaks. Evidence of mating is seen the year round. Gestation is approximately eleven-and-a-half months. The calving interval is not known, though stressed populations off Japan and in the eastern tropical Pacific are mating at earlier ages and calving at shorter intervals than previously, these being typical mammalian responses to overexploitation (similar patterns were reported for Black Sea porpoise stocks that were depleted by overhunting).

Migrations have not been clearly described, though seasonal movements inshore and offshore are indicated. Atlantic spotters move inshore in spring, often approaching close to shore, where they overlap with bottlenose dolphins. Offshore Pacific spotted dolphins may have home ranges of roughly 200 to 300 nautical miles in diameter, but they seem to shift inshore in fall and winter

This spotted dolphin entertained divers on a Bahamian reef during the late 1970s. (Off San Salvador, Bahamas: Al Giddings/Ocean Films Ltd.)

and offshore in spring.

Spotted dolphins are vigorous swimmers, frequently hurling themselves high into the air. They often make a series of amazingly high leaps, fairly hanging in the air for seconds before falling with a splat. When alarmed, herds bunch tightly together to flee.

Food varies with location. Stomachs of offshore eastern tropical Pacific spotted dolphins were found to contain fish of eighteen species and cephalopods of four species; the diet greatly overlapped that of accompanying yellowfin tuna. Atlantic coastal spotters reportedly feed primarily on squid and also take carangid fish, small eels, herring, and anchovies.

Killer whales are suspected predators. Sharks are known to take young animals, and the danger of shark attack no doubt increases during or immediately after tuna purse-seining operations. Confinement in the net and the stress of capture make the dolphins especially vulnerable, and sharks appear to converge in the area of a

Spotted dolphins, conditioned by years of experience, flee after release from a tuna seine in which they were captured briefly. (Eastern tropical Pacific, 11° 14' N, 109° 25' W, October 18, 1976: Donald K. Ljungblad.)

set to take advantage of what, for them, must be a windfall. Spinners and spotters are mentioned in various accounts as victims of attacks by small "blackfish," possibly pilot whales, but more likely (judging from descriptions) pygmy killer whales or melon-headed whales that wait outside purse seines and attack porpoises as they escape from the net.

DISTRIBUTION AND CURRENT STATUS: Spotted dolphins are distributed widely in tropical and some warm temperate waters circumglobally. They have been reported from the tropical South Atlantic and from much of the North Atlantic (New Jersey and England southward, including the Caribbean Sea and the Gulf of Mexico). In the Indian Ocean they range from the Red Sea and the Seychelles east and south to New Zealand, primarily offshore. They are present in the South China Sea and near Japan, and they occur as far south in the Pacific as Peru and New Zealand.

In the eastern Pacific, where they are best studied, spotters are found from the central Gulf of California to about 10° S (coastal); in an area whose corners are at the tip of Baja California, at approximately 10° N, at 145° W, near the Galapagos, and at the northwest coast of Panama; and around Hawaii and throughout the rest of

Polynesia. Spotted dolphins in the eastern tropical Pacific are the main cetacean victims of the controversial tuna fishery in which up to several hundred thousand porpoises were killed each year for some years. Dolphin mortality caused by fishing continues, though it has been reduced, in the United States' fishing fleet that is being monitored; however, it is probably much higher in the growing nonregulated segment of the fishery. By 1979 several million spotted dolphins were believed to remain in the tropical Pacific, but at least one stock had been reduced appreciably by the tuna fishery. The fishery currently is undergoing careful study, and strict regulation is being attempted. In some years, several thousand spotted dolphins are caught in a Japanese shore-drive fishery on the Izu Peninsula, and another drive fishery that takes large numbers of these dolphins has long operated in the Solomon Islands.

CAN BE CONFUSED WITH: Coastal spotters, particularly young, less-spotted animals, closely resemble bottlenose dolphins. They have a slimmer appearance overall, including a generally longer and slimmer beak, and a more complex color pattern, with clearer indications of a cape. In addition, they often have white lips. Adult spotters in many areas probably will not be mistaken, upon close examination, for most bottlenose dolphins, because of the conspicuous spotting on the former. However, some stocks of spotters do not become so heavily spotted, and some bottlenose dolphins have a moderate amount of spotting. North Atlantic observers have noted the presence of a spinal blaze, varying in degree of expression, below and behind the dorsal fin in bottlenose, striped, and spotted dolphins. Care should be taken not to misidentify these animals.

Hump-backed dolphins often are spotted; so in coastal regions of the Indo-Pacific and off West Africa one must take care not to confuse them with spotted dolphins. The peculiarly humped dorsal fin, when present, is probably the most useful characteristic for distinguishing the hump-backed dolphins from spotters.

Striped Dolphin

Stenella coeruleoalba
(Meyen, 1833)
DERIVATION: from the Latin
caeruleus for "sky blue," and
albus for "white.

ZONES 1 TO 5

DISTINCTIVE FEATURES: Black lateral stripes from eye to flipper and eye to anus; white V-shaped "shoulder blaze," originating above and behind the eye, and narrowing to a point below and behind the dorsal fin. Primarily tropical to warm temperate distribution.

DESCRIPTION: This dolphin reaches a length of about 2.7 m. Males are slightly larger than females. Sexual maturity is reached at about 1.8 to 1.9 m. Average length at birth is about 1 m.

The striped dolphin's beak is similar in shape and proportion to those of the spotted, common, and bottlenose dolphins. The forehead is not prominent, and slopes smoothly from beak to blowhole, but with a distinct crease separating forehead and beak. The moderately falcate dorsal fin is at the middle of the back. The caudal peduncle is narrow without a strong keel.

This slender dolphin's coloration pattern is the key to recognizing it at sea. In general, its dorsal color varies from light gray to dark gray to bluish gray; its sides are light gray; the belly is white. A black band begins behind the eye, and extends along the flank to the anus. Another small black stripe originates at the same point near the eye, and begins to run parallel and almost adjacent to the first band. However, this stripe soon bends toward the flipper and fades into the whiteness of the flank just above the flipper. Another broader stripe starts below and in front of the eye, and extends to the anterior origin of the flipper. Some investigators have used the differences in this eye-to-flipper stripe (sometimes it is double) to describe two species of striped dolphin.

In addition to these distinctive black stripes, most animals have what appears to be a finger of light pigmentation intruding into the basically dark cape. This light spinal blaze originates as an open V above and behind the eye, narrowing to its vertex below and behind the dorsal fin. Although many spotted dolphins and bottlenose dolphins in the North Atlantic have a similar blaze, that of the striped dolphin is generally bolder and more sharply defined.

There are 45-50 pairs of sharp, slightly incurved teeth in the upper and lower jaws.

NATURAL HISTORY: The striped dolphin is gregarious; it is commonly found in aggregations numbering a few hundred, and sometimes in herds of several thousand (2,838 were once captured from a single herd off Japan's Izu Peninsula). There is marked segregation by age and sex among striped dolphins. Large schools that contain primarily subadults can be completely devoid of adult males, and have only a token number of mature females.

In the western North Pacific, where it has been most thoroughly studied, the striped dolphin has a prolonged breeding season, with apparent peaks of mating activity in winter, spring, and possibly late summer. Gestation lasts for about twelve months. Some young animals begin eating solid food three months after birth, but weaning is often not completed until well into the second year. Age at sexual maturity may be five to six years, or it

The handsomely painted striped dolphin, a common sight on the tropical and subtemperate seascape. (Southwest of Sri Lanka, March 1982: Abigail Alling – WWF.)

may be nine. Mature females probably bear a single calf every three years.

In Japan a migratory pattern exists. Striped dolphins approach the coast in September and October, and move southward along the coast, apparently dispersing into the East China Sea for the winter. In April they return along roughly the same path, but farther offshore. They eventually leave the coast to summer in the pelagic North Pacific.

These are active and conspicuous creatures. They frequently jump clear of the water and are known to ride bow waves. Strandings of individuals are not uncommon, but we are unaware of any mass strandings. This species has not been successfully maintained in captivity.

Diet includes a variety of small mesopelagic fishes, as well as shrimp and squid. Predators have not been identified, although sharks and killer whales are likely.

DISTRIBUTION AND CURRENT STATUS: The striped dolphin is widely distributed across all temperate, subtropical, and tropical seas. So far, only a few discrete stocks have been proposed, including one off South Africa, one or two in the eastern tropical Pacific, and another in the western North Pacific. It is said to be the most common cetacean in the Mediterranean Sea. Because of its usual offshore distribution, little has been learned outside Japan about its biology.

Fishermen on the Pacific coast of Japan have hunted striped dolphins for at least several centuries, with hand harpoons and by driving them ashore. Annual catches during the 1960s ranged as high as 20,000 animals. In other areas, such as Papua New Guinea and the Solomon Islands, they are included in native catches of small cetaceans. Because yellowfin tuna sometimes associate with them, these dolphins experience some incidental mortality in the eastern tropical Pacific purse-seine fishery.

The western North Pacific population hunted off Japan has been estimated as 400,000. The most recent estimate of their numbers in the tropical Pacific is several hundred thousand. There are no estimates for other areas.

CAN BE CONFUSED WITH: There is an excellent chance that striped dolphins may be confused with common dolphins when they are sighted at sea. Although in most stocks striped dolphins are somewhat larger than common dolphins (long-beaked coastal common dolphins in some areas of the eastern tropical Pacific are larger than nearby stocks of striped dolphins), the only way to distinguish the two in most areas is by close attention to color patterns. The common dolphin lacks the distinctive stripes and blaze of the striped dolphin. Instead it has its own unmistakable hourglass or crisscross effect on the flanks, together with a black stripe from the flipper to the middle of the lower jaw.

At least in the North Atlantic, striped dolphins can be confused with spotted and bottlenose dolphins because of the presence of a spinal or shoulder blaze. Adult spotted dolphins in this region, however, are likely to be heavily spotted. In bottlenose dolphins the blaze is usually muted and less well-defined than it is in the striped dolphin. If the sides of the animal are clearly seen, the eye-to-anus stripe of the striped dolphin is diagnostic. The striped dolphin might be confused with Fraser's dolphin in the tropics, since the latter has broad, alternately light and dark diagonal stripes on the sides. However, Fraser's dolphin has a short beak, short flippers, and a very modest triangular dorsal fin.

Long-snouted Spinner Dolphin*

Stenella longirostris
(Gray, 1828)
DERIVATION: from the Latin *longus* for "long," and *rostrum* for "beak, snout."

ZONES 1 TO 5

Short-snouted Spinner (or Clymene) Dolphin*

Stenella clymene
(Gray, 1850)
DERIVATION: from the Greek *Clymene,* daughter of Oceanus and Tethys, mother of Phaeton by Apollo.

STENELLA LONGIROSTRIS

STENELLA CLYMENE

DISTINCTIVE FEATURES: Long, slim snout (*longirostris* only); lips and tip of snout dark; habitually jumping and spinning on longitudinal axis; dorsal fin usually triangular or canted slightly forward, often with a light spot in center of dark area; tail stock of adult males usually strongly keeled; distribution tropical to warm temperate, primarily in deep water.

DESCRIPTION: Five forms of spinner dolphins have been described. Recent understanding of all but one is based mostly on eastern tropical Pacific specimens. The largest form, the Costa Rican spin-

Bubbles escape from the blowhole of a whistling eastern spinner dolphin, trapped in a tuna purse seine. (Eastern tropical Pacific: courtesy of American Tunaboat Association.)

ner (males to about 2.2 m long; females, about 2.1 m), is long, slender, and all gray; the males have a triangular or forward-canting dorsal fin. The similarly shaped and colored eastern spinner is shorter (males to about 1.9 m; females, 1.8 m), and it is distinguishable from the Costa Rican form by skull differences. The more pelagic whitebelly spinner is larger and more "girthy" (to about 2 m long), and it has some white on the belly, with sharper color differentiation from the back and sides to the belly. Its dorsal fin tends to be falcate more often. Intergrades between the eastern and whitebelly forms are seen occasionally. The Hawaiian form is similar to the whitebelly but larger (to over 2 m), with a still more complex

---

* For several species of smaller cetaceans currently under detailed study, scientific names and preferred common names are subject to change as understanding improves.

color pattern (consisting of three elements) and a consistently fal-cate dorsal fin. The fifth form, the so-called short-snouted spinner (*S. clymene*), most closely resembles the Hawaiian form in shape and color, but, as the common name implies, it has a shorter snout. It is also more robust than the long-snouted spinners.

Newborn eastern and whitebelly spinners are 70 to 85 (average 77) cm long. There is little appreciable difference in size at birth in other populations.

In general, the body and beak of spinner dolphins (except *S. clymene*) are long and slender. The head slopes gently toward the snout, and the forehead is not bluff. Nevertheless, there is a definite crease at the apex of the melon. The flippers of adults, at least, are proportionately longer than those of spotted dolphins.

As noted, the dorsal fin differs in shape between stocks, but in general it tends to become more erect with age in each population, reaching its most exaggerated proportions in adult males.

The color pattern includes three basic elements, which to-gether may give a dark gray color to the back, a lighter tan cast to the sides, and an even lighter, often white, appearance to the belly. In the Costa Rican and eastern forms, the elements are obscured, and the body appears almost uniformly battleship gray. In whitebellies the ventrum is variably white, but demarcation of colors on the sides and back is largely lacking. In the Hawaiian and short-snouted varieties, all color pattern elements are well-defined and clearly visible. In spinners with complex coloration, the dorsal fin is often light in the center, dark on the margins.

There are 45-65 or more sharply pointed teeth in each row in long-snouted spinners. Short-snouted spinners have fewer teeth (38-49 per row).

NATURAL HISTORY: Spinner dolphins may occur in herds of over 1,000 animals, though herds of 200 or fewer are common. They fre-quently associate with spotted dolphins in the eastern tropical Pacific, and with other oceanic dolphins and small to medium-size whales (e.g. pilot whales, pygmy killer whales, and melon-headed whales) in much of their range. Short-snouted spinners have been observed off the coast of West Africa in a large herd of common dolphins.

The most complete reproductive data are available for eastern Pacific spinners. Males reach sexual maturity by about 1.7 m, females by about 1.65 m. In the larger (and unexploited) form in-habiting the Gulf of Mexico, sexual maturity is reached at a some-what greater size in both sexes, about 1.9 m and 55 to 60 kg. Age at sexual maturity for males is similar in both populations (ten to twelve growth layers), but female eastern spinners appear to mature more quickly (at a mean of 5.5 layers) than females in the Gulf of Mexico (seven to ten layers). In unexploited populations, adult females give birth to a single calf every second or third year, after an

Three Hawaiian spinner dolphins form a high-spirited silhouette. (Off Hawaii: Bernd Würsig.)

Without the ever-playful spinner dolphins, the high seas would seem so much less alive. (Southwest of Sri Lanka, March 1982: Abigail Alling–WWF.)

average 10.6-month gestation period; birth is more frequent in depleted populations. The calf is weaned at an age greater than seven months. There is no evidence of simultaneous pregnancy and lactation.

Except for the coastal form near Hawaii, the Costa Rican form, and the Gulf of Mexico population, all of which approach close to shore at least occasionally, spinners are offshore, deepwater dolphins. Strandings, sometimes involving several tens of individuals, occur from time to time.

In the Pacific, spinner dolphins frequently associate with yellowfin tuna, less frequently with skipjack tuna, as part of a still incompletely understood but tight bond. The high-seas tuna fishery exploits this association. The porpoises, and the fish accompanying them, can be herded by high speed skiffs, then encircled with huge seine nets. The bag created by the pursing of the net leaves both fish and porpoises trapped. The high-strung, pelagic spinners are not always released successfully from the net; so sometimes the mortality is very high.

Spinners are animated and acrobatic, often jumping from the water and spinning repeatedly on their longitudinal axis, a spectacular act they may repeat many times in a short period of apparent exuberance. Though sounds caused by their splashing re-entry may be used for communication, this behavior often looks like sheer play. Spinners in many areas are willing bow riders, although those in heavily exploited portions of the tropical Pacific now flee from, rather than approach, ships.

Food of spinner dolphins in the eastern tropical Pacific is small, mainly mesopelagic fish and epipelagic and mesopelagic squid. The short-snouted spinner dolphin appears to be a midwater or night feeder that preys on small fish and squid.

Although they are not common display animals, spinners have been maintained in captivity for several years at a time in Hawaii.

DISTRIBUTION AND CURRENT STATUS: Spinner dolphins are found in the Atlantic, Indian, and Pacific oceans, where they are restricted to tropical, subtropical, and, less often, warm temperate regions. Primary distribution is in pelagic zones, though they venture into shelf waters off western Central America and the southeastern United States. Specimens have been collected from the waters near various South Pacific islands, Australia, the Solomon Islands, New Guinea, Indonesia, Japan, Ceylon, Madagascar, eastern and western Africa, the Caribbean Sea, the coast of the eastern United States, and the Gulf of Mexico.

Different distributions have been attributed to each identified form in the eastern tropical Pacific. The Costa Rican spinner is found primarily less than 150 km from shore along the western coast of Central America, between about 14° N and 6° N. Eastern spinners range from the southwestern coast of Baja California south to the equator and offshore to about 126° W. Whitebellies occupy much of the equatorial Pacific, well offshore, extending south to 14° S off western South America, north to 17° N due west of Mexico, and west almost to the Hawaiian Islands. Eastern and whitebelly spinners greatly overlap in range, and are sometimes found in mixed herds.

In the Atlantic, their presence has been confirmed between Cape Hatteras and Rio de Janeiro in the west, and between the equator and 20° N off the western African coast.

Short-snouted spinners are known to inhabit the tropical and subtropical Atlantic only, with records from the southeastern United States (actually as far north as New Jersey), the Gulf of Mexico, the Caribbean Sea, the northwestern coast of Africa, and the mid-Atlantic.

Drive fisheries in the Solomon Islands and Japan are known to involve spinners at least occasionally, and the local harpoon fishery at St. Vincent in the Lesser Antilles has been known to catch them.

The status of stocks exploited by the tuna fishery is assessed periodically in order to set allowable catch levels. Eastern spinners are the most seriously depleted; the northern stock of whitebellies has also been hard-hit by the tuna industry. The total number of spinners in the eastern Pacific is probably a few hundred thousand; there probably were at least a million before purse seining for tuna began in 1959.

CAN BE CONFUSED WITH: Costa Rican and eastern spinners should be distinguishable from other oceanic dolphins by their seemingly uniform battleship-gray coloration and their long snouts. The presence in most groups of at least a few individuals with forward-canted dorsal fins and thick keels should also help distinguish spinners in many areas.

As their color pattern becomes more complex, spinners might be mistaken for bottlenose dolphins or common dolphins. In these last two species, the beak is not as long and narrow as that of the long-snouted spinner, but is similar to that of the short-snouted spinner. Spinner dolphins have black lips and a black-tipped beak, but the other two have gray or white-tipped beaks. The common dolphin has a very distinctive color pattern, the cape forming a dark V-shaped intrusion on the sides below the dorsal fin, whereas the cape of spinner dolphins assumes a sinuate shape below the dorsal fin.

Southern Right Whale Dolphin

Lissodelphis peronii
(Lacépède, 1804)
DERIVATION: from the Greek
lisso for "smooth," and *delphis* for
"dolphin"; F. Peron was a
French naturalist who observed

these dolphins at 44° S off
Tasmania.

ZONES 3 TO 5, NORTHERN
FRINGES OF 6

DISTINCTIVE FEATURES: No dorsal fin; body slim and graceful; striking black and white coloration; white zones covering head and extending onto sides, including flippers, and onto tail stock; small but distinct beak; limited to temperate portions of the southern hemisphere.

DESCRIPTION: The few specimens that have been examined were up to about 2.35 m (females) and 2.3 m (males) long. Free-swimming animals off South Africa have been estimated as about 3 m. The body is generally slim and has been described by one author as more broad than deep in the thoracic region; this characteristic might help stabilize the dolphin, which lacks a dorsal fin. Though short, the beak is distinct and clearly demarcated from the fore-head. The lower jaw sometimes extends past the tip of the upper jaw. The few specimens examined have had 44–49 pairs of small, sharp teeth.

  These finless right whale dolphins derive their common name from the bulky right whale, which also lacks a dorsal fin. The strik-ing white and black zones may differ in extent and intensity, but are separated by a crisp border. Usually the white covers the entire ventral surface, spreading up the sides to include most of the head in front of the blowhole and the flanks. The flippers have various amounts of white, and patterns within a herd are thought useful for reidentifications of a herd on repeat sightings. The dorsal surface of the tail flukes is dark gray, paling to a white leading edge. Animals with all-black, all-white, and other anomalous coloration have been observed. Of the coloration of the head of this species, which he called the mealy-mouthed porpoise, Melville wrote "the white comprises part of the head and the whole of his mouth, which

Mealy-mouthed porpoises streak to the surface for a breath. (Off South Africa, about 27° S, 14° E, May 1979: Roy Cruickshank.)

makes him look as if he had just escaped from a felonious visit to a meal bag."

NATURAL HISTORY: Though the number of sightings and strandings are increasing as fishing and research increase in pelagic regions of the temperate southern hemisphere, little is known about the southern right whale dolphin's natural history. The few dozen groups reported have contained from two to more than a thousand animals. The species occurs outside the 100-fathom curve near New Zealand, and in deep offshore waters off South Africa, but has been described as both coastal and oceanic off Chile. It is a fast swimmer, reaching at least 25 km per hour, and often lobtails and leaps clear of the water. Entire herds so moving have been described as traveling in "a series of bouncing leaps." Individuals and small groups occasionally bow ride. Like the northern right whale dolphin, the southern right whale dolphin frequently is found associating with the dusky dolphin, the southern-hemisphere counterpart of the Pacific white-sided dolphin, and with pilot whales.

From stomach contents (off Chile and New Zealand) and net samples near feeding animals (off South Africa), it is known to feed on myctophids and squid.

DISTRIBUTION AND CURRENT STATUS: This is a circumpolar, marginal Antarctic species that remains almost exclusively in temperate waters. Although there are a few records south of the Antarctic Convergence, most are more northerly, in the West Wind Drift. It ranges in the Humboldt Current north as far as 19° S off western South America, and has commonly been reported seaward of the 100-fathom curve off New Zealand, in the Falkland Current, offshore from South Africa, and in coastal and oceanic habitats to 450 km from the coast of Chile. Seasonal patterns of distribution are unknown, though South African records are for summer.

Southern right whale dolphins reportedly were taken sporadically by whalers in the nineteenth century, and they continue to die occasionally from incidental hooking or entanglement in nets of fishermen off Chile. Current stock sizes are unknown, and there is now no deliberate fishery for them.

CAN BE CONFUSED WITH: Because it lacks a dorsal fin, this handsome, strikingly marked dolphin is unlikely to be mistaken for any other cetacean, though it does resemble fur seals and other pinnipeds when it is swimming slowly.

Northern
Right Whale Dolphin

Lissodelphis borealis
(Peale, 1848)
DERIVATION: from the Latin
borealis for "northern."

ZONE I

DISTINCTIVE FEATURES: No dorsal fin; body slim and graceful; coloration visible in water primarily black; small amounts of white ventral hourglass pattern visible around flippers; small but distinct beak; limited to temperate North Pacific.

DESCRIPTION: Males grow to at least 3.1 m, females 2.3 m. Males are sexually mature by 2.2 m, females by about 2 m. Live newborn have been estimated to be about 80 to 100 cm long.

The body is generally slim, particularly in the region of the tail stock, and is most recognizable by the total absence of a dorsal fin or any trace of a dorsal ridge. The flippers are slim, sharply pointed on the tip, and on the posterior margin marked by two concave lines joining to form a point near the middle. The flukes are surprisingly narrow for such a rapid swimmer and are gracefully curved and pointed on the tips.

The melon slopes gently forward into a small but distinct beak, which, as on its southern counterpart, is clearly demarcated from the forehead by a transverse groove. The body is basically black, but with a striking white lanceolate pattern of varying extent on the ventral surface. These vivid white markings extend anteriorly from the caudal peduncle as a narrow band (apparently narrower in males than in females) before expanding into a broad ventral patch which reaches the axilla, then converges toward the gular region. There is characteristically a small white mark at the tip of the lower jaw. The flippers are all dark. The flukes are light gray dorsally; ventrally the distal portions are white. Newborn seen at sea have been reported to appear cream to light gray, but juveniles assume the coloration of adults during the first year. There are about 36–49 teeth per row.

NATURAL HISTORY: The species occurs in very large herds of up to several thousand. Average sizes of about 200 and 110 have been reported for the western and eastern North Pacific, respectively. In all areas they associate with a variety of other cetacean species, especially Pacific white-sided dolphins, with which they share an extensive common range.

Newborn are reported most frequently in early spring. Little else is known about the species' reproduction.

In both the western and eastern North Pacific, the species' distribution appears to shift south and inshore between October and May or June, then north and offshore again from summer through fall.

Right whale dolphins may reach speeds greater than 40 km per hour in bursts, and appear to sustain speeds of 25 km per hour easily for more than thirty minutes.

Though they sometimes ride bow waves, the species more often shuns human contact and runs from boats. Two escape modes have been identified. In the first, the dolphin barely breaks the surface to breathe, an effective escape for the finless dolphin in all but the flattest seas. If this behavior is ineffective, the herd often breaks into a series of low-angle leaps, each covering as much as 7 m, bunching closely as it runs. A herd escaping in this manner will create a sizeable disturbance on the water surface. Right whale dolphins are always graceful and fleet.

At least 17 fish species have been identified from stomach contents, the most common of which are myctophids and bathylagids. Squid, however, is the most commonly observed food item, and migrations have been thought to be related to movements and availability of spawning squid.

DISTRIBUTION AND CURRENT STATUS: Northern right whale dolphins are restricted to the North Pacific, where they appear to be widely distributed in a crescent-shaped region corresponding to moderate temperate currents, apparently not entering tropical waters, and rarely entering subarctic or the coldest temperate waters. In the western North Pacific they are known from Cape Najima, northern Honshu, and Cape Inubo, Japan, north to about 51° N. They are reportedly common in the northern Sea of

The low-angle leaps typical of a large running herd of northern right whale dolphins, among the most unexpected of dolphin forms. (Off Point Conception, California, October 1979: Robert L. Pitman.)

Japan and very common off the Pacific coast of Japan, particularly to the north. From Japan the distribution gradually tapers east-northeast. In the eastern North Pacific, they range from British Columbia (approximately 50° N) to about the Baja California border (32° N), with stragglers to 29° N. Most records are from California, though this may well reflect a more offshore distribution off the less heavily populated Oregon, Washington, and Canadian coasts. Eastern and western Pacific populations may be reproductively isolated by an area of very low density south of the western Aleutians.

The species appears everywhere to favor deeper-water habitats, but does approach shore at the heads of deep canyons, particularly in winter months. On both sides of the Pacific, populations shift southward from October until as late as May or June, then shift northward.

Northern right whale dolphins have been the victims of coastal shore fisheries off Japan, and are occasionally entangled in North Pacific gill-net and seining operations. Small numbers have been captured live off California for aquarium use. They are among the most abundant of the oceanic dolphins that inhabit the temperate zone of the North Pacific.

CAN BE CONFUSED WITH: As the only dolphin species in their respective hemispheres that lack a dorsal fin, northern and southern right whale dolphins should, upon close examination, be readily distinguishable from all other dolphin species. But when rapidly fleeing or swimming slowly (hence creating little surface disturbance), right whale dolphins and various pinnipeds, primarily fur seals and sea lions, might be mistaken for one another.

Heaviside's Dolphin*

Cephalorhynchus heavisidii
(Gray, 1828)

DERIVATION: from the Greek *kephale* for "head," and *rhynchos* for "nose, snout," because the rostrum, which is about half the length of the skull, is well-differentiated from the rest of the head; Captain Haviside (the first "e" has been added in error) was the employee of the British East India Company who conveyed the type specimen from the Cape of Good Hope to England in 1827.

ZONE 5

DISTINCTIVE FEATURES: Small size; no beak; triangular dorsal fin; all black dorsally, with three lobes of white on each side originating on belly; limited to coastal waters of southwestern Africa.

DESCRIPTION: As far as is known, average adult length is about 1.2 to 1.4 m. The head lacks a distinct beak, and the forehead slopes gently toward the snout. The dorsal fin is broadly triangular and blunt at the peak; the flippers are ovate. The trailing edge of the flukes is strongly concave, and there is a median notch.

Because no photos were available to us, this species was painted to resemble other members of the genus according to written descriptions of it.

Coloration is distinctive: black dorsally and white ventrally. The black and white are sharply demarcated. Three lobes of white extend from the ventrum onto each side, one fore and one aft of

---

*This dolphin was not painted from photographs (few, if any, exist), but from descriptions and historical sketches.

the flipper, and an especially prominent one below and slightly behind the dorsal fin, swept obliquely toward the tail. There are 25-30 pairs of small, pointed teeth in the upper and lower jaws.

NATURAL HISTORY: Very little is known about the biology and behavior of this coastal species. Investigators have found that the sounds made by this and the other species of *Cephalorhynchus* closely resemble those of *Phocoena*. The two genera have numerous morphological characteristics in common (lack of a beak, small rounded flippers, and a low dorsal fin). In addition, both are primarily coastal in distribution, and are generally shy and difficult to approach. Heaviside's dolphin is believed to feed on squid and bottom-dwelling fish.

DISTRIBUTION AND CURRENT STATUS: Range is limited to the Benguela Current system off southwestern Africa, from the Cape of Good Hope northward to about 18° S, apparently only in coastal waters.

Although no direct hunting is known to occur, this species is vulnerable to accidental capture in purse seines. Less than 100 are believed to be captured each year by the extensive commercial purse-seine fishery for pelagic fish off South Africa and southwestern Africa. Also, a few may be caught in beach seines from time to time.

Hector's Dolphin

Cephalorhynchus hectori
(van Beneden, 1881)
DERIVATION: Hector was a New
Zealand zoologist who first
collected the species in 1869.

ZONE 3

DISTINCTIVE FEATURES: Rounded dorsal fin with convex rear margin; no beak; black appendages and mask; light forehead; found only in coastal waters around New Zealand.

DESCRIPTION: This is a small dolphin, with a maximum length of about 1.6 m and an average adult length of about 1.2 to 1.4 m. Alan Baker, who has studied it in some detail, has described Hector's dolphin as "dumpy" in appearance, noting that its girth can be as much as 68 percent of total length.

It has no beak; the mouth slants up toward the eyes. Flippers are small and rounded at the tips; the trailing edge of the flukes is concave, with a shallow median notch. The dorsal fin is rounded, with a strongly convex posterior margin in adults, and a notch at the posterior base.

It is distinctively colored, but the pattern is complex. The sides of the head, flippers, dorsal fin, and tail (beginning well ahead of the flukes) are all black. The very tip of the mandibles is black, but the throat and most of the lower jaw, as well as the belly, are white. The black of the face is continuous with that of the flippers and continues in a line on the ventrum, making a V at the midline, pointing tailward. A thin black line curves over the head behind the blowhole. The forehead anterior to this black line is gray ("finely streaked with black"), and the remainder of the animal is pale gray, except for a "finger" of white that extends from near the genital region onto the sides, pointing toward the tail, and a white patch behind each flipper. Differences of color pattern between live and dead specimens, resulting from post-mortem changes, have led to considerable taxonomic confusion.

There are 26–32 pairs of small teeth in upper and lower jaws.

A Hector's dolphin just before capture and tagging by a research team studying the dolphins' movement patterns. (Cloudy Bay, New Zealand, December 1978: Alan N. Baker.)

NATURAL HISTORY: Little is known about this inshore species. Groups are small, seldom more than two to eight animals. Occasionally groups of more than twenty are seen. Their movements appear to be local, and they probably do not undertake extensive migrations. River mouths, particularly when muddy and discolored, seem to be favored habitats.

Hector's dolphins bow ride, more often in the wake than the bow of a vessel. They are apparently unable to remain for long with a boat traveling at more than 18 km per hour. These dolphins are difficult to observe, since they show little of themselves when surfacing, and they rarely jump. They do, however, surface frequently, often at intervals of less than half a minute.

The few stomach contents examined show that these dolphins have a varied diet, including shellfish, crustaceans, small fish (such as red cod, horse mackerel, anchovies, flounders, and sand star-gazers), and squid.

Several have been taken alive and maintained in an oceanarium in New Zealand, and have proved amenable to training.

DISTRIBUTION AND CURRENT STATUS: Hector's dolphin inhabits coastal waters of New Zealand between Bay of Islands (35°15′ S) and Foveaux Strait (46°35′ S). It is especially common along the northeastern coast of the South Island. Usually it is found within 8 km of shore, and in water less than 80 m deep. The Banks Peninsula area and Cloudy Bay are said to be especially good spots for sighting this species. Some are also seen off Westport.

A few have been taken alive for display, and some are caught accidentally in nets; otherwise the species has never been exploited on a significant scale. Its inshore distribution suggests that it may be susceptible to various forms of environmental disturbance and to much incidental mortality in nets.

CAN BE CONFUSED WITH: Because of its limited distribution and striking coloration, Hector's dolphin should not be confused with any other species in the New Zealand region. It does, however, bear a resemblance to the black dolphin (*C. eutropia*) of South America.

Black Dolphin

Cephalorhynchus eutropia
(Gray, 1846)
DERIVATION: from the Greek *eu*
for "well, good," and *tropis* for
"head" or *tropidos* for "keel,"
together referring to the skull,
described as "strongly keeled in
the center behind."   ZONE 4

DISTINCTIVE FEATURES: Low, rounded dorsal fin; no beak; dark pigmentation; limited to the coastal waters of Chile.

DESCRIPTION: Adult body length is said to be about 1.6 m. This species is not well-described in the literature. Presumably it is similar in body size and shape to the better-known members of the genus. Coloration is dark, with white throat and belly and small white patches or spots behind the flippers. The top of the head is marked by a light-gray region from the blowhole to the snout tip. There have been 30–31 pairs of teeth in the upper and lower jaws of specimens examined and reported in the literature.

NATURAL HISTORY: Nothing is known about its biology. It is a very shy, coastal species, found in groups of eight to fourteen individuals.

DISTRIBUTION AND CURRENT STATUS: Known range includes only coastal waters of Chile, from approximately Concepción (37° S) southward through the fjord region to Navarino Island (55° S), near

The black (Chilean) dolphin is found only in the coastal waters of southwestern South America. (Corral, Chile: Kenneth S. Norris.)

Cape Horn, and including the channels of Tierra del Fuego. Fishermen in some parts of Chile are known to use this dolphin for bait and possibly for human consumption. The dolphins are taken incidentally in gill nets and, rarely, by harpoon. Extent of this exploitation is unknown.

CAN BE CONFUSED WITH: Although its range overlaps that of *C. commersonii,* there is little likelihood of confusing these two species because of the latter's striking black and white coloration. In the northern part of its range, *C. eutropia* is very likely to be confused with Burmeister's porpoise. The best external distinction between the two is probably dorsal-fin shape: Burmeister's porpoise has a triangular fin, with a more nearly pointed tip and knobs on its anterior border.

Commerson's Dolphin

Cephalorhynchus commersonii
(Lacépède, 1804)
DERIVATION: Commerson was
an eighteenth-century French
medical doctor and botanist
whose description of the species
from observations near the

Strait of Magellan was used by
Lacépède.

ZONES 3 AND 5

DISTINCTIVE FEATURES: Rounded dorsal fin; unique color pattern;
distribution limited to western South Atlantic and near southern
tip of South America; also known from vicinity of Kerguelen Island
in the southern Indian Ocean, primarily in coastal waters.

DESCRIPTION: Maximum body length is about 1.7 m. The body is
thick and stocky, with a sloping forehead and no beak. The mouth
is small and straight, slanting up toward the eye. The elliptical
flippers are rounded distally, and the trailing edge of the broad
flukes is concave, with a small median notch. The most distinctive
character other than coloration is the rounded dorsal fin, set
slightly behind the middle of the back. The tip is broadly rounded
and the rear edge slightly concave. Coloration is striking. The head
is black, with a white chevron on the throat. The black coloration
ends at approximately the neckline dorsally, but slants downward
and backward to the all-black flippers, making the blackness continuous ventrolaterally between the snout and flippers. A black
band connects the flippers ventrally, and the black coloration on
both dorsum and ventrum ends in a point at the respective longitu-

Visitors to the Falkland Islands in the last decade report Commerson's dolphins to be common in and near the major harbors. (Port Stanley, Falkland Islands, January 1974: Frank S. Todd, Sea World, Inc.)

dinal midlines. A wide black stripe begins in front of the all-black dorsal fin and continues along the spine to the middle of the tail stock, posterior to which the entire animal is black. Except for a black genital patch, the rest of the animal is white. The white and black zones are sharply demarcated in adults. Newborn calves are said to be completely brown; the handsome black and white markings develop with age.

There are usually 29-30 pairs of small, pointed teeth in the upper and lower jaws.

NATURAL HISTORY: Very little is known about the natural history of Commerson's dolphin. It usually occurs in shallow coastal waters in small groups of two to twelve. The largest group seen in an Argentine harbor numbered 31 animals, of which 11 were calves. No other member of its genus is found around offshore islands.

Near-term fetuses have been found in Commerson's dolphins at Tierra del Fuego during December, suggesting that the peak season for calving may be early in the austral summer.

An Argentine naturalist, Jorge Mermoz, made a series of observations of Commerson's dolphins in a shallow harbor. He estimated 6 to 7 knots as their usual swimming speed, 15 to 20 seconds as their usual diving time. While submerged, their movements were erratic, making it difficult to anticipate where they would next surface. Mermoz was impressed by how frequently they leaped. He counted 65 to 70 breaches by six individuals within a period of 17 minutes.

The only examined stomach contents included krill and remains of squid, crab, and "cuttlefish." Mermoz believes they feed on Fueguian sardines, silversides, and southern anchovies in San Jorge Gulf, where he made his observations.

DISTRIBUTION AND CURRENT STATUS: Commerson's dolphin is known only from the east coast of South America between Peninsula Valdes (about 42° S) and southern Tierra del Fuego, where it is very common, from Magellan Strait, from Drake Passage (between Tierra del Fuego and the Antarctic Peninsula), from the coast of Chile south of 50° S, and from around the Falkland Islands, South Georgia, and Kerguelen Island (in the southern Indian Ocean). Sightings are most frequent in Magellan Strait and around the Falklands.

A few are killed accidentally in gill nets set for crabs along the Argentine coast, and local fishermen may kill them occasionally for food. Several have been shipped to aquariums abroad, and the prospect exists for the development of a regular live-capture fishery based in Argentina. Commerson's dolphins inhabit nearshore waters, including heavily trafficked harbors where there is potential for conflict with human activities.

CAN BE CONFUSED WITH: Attention to coloration should enable the careful observer to avoid confusing this with other species. The color pattern and body shape strikingly resemble that of *Phocoenoides dalli truei*, but the two species live in different hemispheres.

Such companionship could not occur in nature. The Arctic beluga and the subantarctic Commerson's dolphin are normally worlds apart. (Duisberg Zoo, Duisberg, West Germany: W. Gewalt.)

8. The True Porpoises

Family: Phocoenidae

Harbor Porpoise

Phocoena phocoena
(Linnaeus, 1758)
DERIVATION: from the Greek
phokaina or the Latin *phocaena*
for "porpoise."

ZONES 1, 2, 7, AND 8

DISTINCTIVE FEATURES: No beak; triangular dorsal fin; lack of sharply defined pigmentation pattern; generally undemonstrative, and wary of or indifferent to boats; distribution primarily coastal, limited to temperate and subarctic northern hemisphere.

DESCRIPTION: Maximum size is just over 2 m and 90 kg, but average adult size is nearer 1.5 to 1.6 m and 45 to 60 kg. Length at birth is usually between 70 and 90 cm.
  This animal is small and chunky. There is no forehead or snout;

the short, straight mouthline tilts slightly upward.

The flippers are short and not sharply pointed. The triangular dorsal fin is slightly aft of midbody, and often is not sharply pointed. It is usually in the shape of a triangle, with the posterior edge vertical or slightly tilted forward (it may even be slightly falcate in some cases). Occasionally a series of blunt spines (bumps) is present on the anterior margin of the flipper or dorsal fin. The tail stock is flattened laterally into a noticeable keel; the flukes are separated by a median notch, and the trailing edge is moderately concave.

Coloration is variable and is not cleanly patterned. It is usually

Such an expression would never be seen in nature, where harbor porpoises maintain a businesslike approach to life. (Harderwijk Aquarium, The Netherlands: Gerard Van Leusden.)

dark brown or gray on the back, fading to lighter grayish brown on the sides, often with speckling in the transition zone and white on the belly. A grayish white zone usually extends farther up the sides in front of the dorsal fin. This lightened zone can be seen as the animal arches at the surface to breathe. The dark appendages and dark, narrow stripe(s) between eye and flipper can usually be seen only on beached animals. Albinos and partial albinos have been reported.

There are 19-28 small, spade-shaped teeth on each side of the upper and lower jaws.

NATURAL HISTORY: Though adventitious aggregations of fifty or more harbor porpoises are seen occasionally, they are more often encountered singly, in pairs, or in groups of five to ten. The larger concentrations may simply be loose coalitions of these small groups, formed seasonally for migration or for taking advantage of especially rich feeding grounds.

There is much diversity of opinion about the species' reproductive biology. The principal mating season is reportedly summer, from June to possibly as late as October. Gestation probably lasts about 11 months. Some births have been reported as early as March, but the majority occur between May and July. The period of calf dependency is still unknown, as is age at sexual maturity; reasonable guesses might be six to eight months and three or four years, respectively. Some calves are known to have taken significant quantities of solid food at an age of three to four months. Annual calving has been inferred in the Bay of Fundy, possibly in response to population reduction or food availability, but a two-year interval may be more usual. Life span is relatively short, perhaps not exceeding 15 years.

Most seasonal movements seem to be inshore-offshore rather than north-south. These may be determined by availability of food or ice-free water.

The harbor porpoise does not come to vessels to bow ride; in fact, it is often difficult to approach. Breaching or porpoising clear of the water is infrequently observed in most of the species' range, although, when chasing prey, these porpoises do make arc-shaped leaps above the surface and sometimes considerable splashing occurs as they race along the surface. When on the move, they surface frequently to breathe (as many as six or eight times at roughly one-minute intervals). When feeding, they surface less often (three or four breaths at two- or three-minute intervals). When pursued, they can swim at speeds approaching 22 km per hour. The few captive specimens that have survived for months or a few years have allowed some study of their sounds. They are known to echolocate.

The harbor porpoise eats a wide variety of fish and cephalopods, but seems to prefer schooling, nonspiny fish, such as herring, mackerel, sardines, pollack, and whiting.

Great white sharks and killer whales are known predators, but the frequency of predation is largely unknown. In the Bay of Fundy, where much research has been done on the harbor porpoise, great white sharks are common enough to be regarded as significant predators on these small and vulnerable cetaceans. Ice entrapment has been known to cause mass mortality.

DISTRIBUTION AND CURRENT STATUS: In general, this is a coastal species, limited to cold temperate and subarctic waters of the northern hemisphere. It is frequently found in bays, rivers, estuaries, and tidal channels of western Europe and along both coasts of North America. Along the Pacific coast of Asia, it is

A "puffing pig," as the harbor porpoise is known locally in much of eastern Canada and New England, removed from a herring weir, this time to be tagged and released. (Passamaquoddy Bay, eastern Canada, August 26, 1980: Randall R. Reeves.)

present but not abundant. An isolated population inhabits the southern Sea of Azov and the Black Sea.

In the eastern Atlantic it occasionally reaches as far north and east as Novaya Zemlya and the Kara Sea, becoming more abundant near Murmansk and Varanger Fjord. It penetrates deep into the Baltic Sea (even to Lake Ladoga and the Gulf of Bothnia). It is not known for certain whether they enter the Mediterranean. Specimens collected at Senegal, West Africa, represent the most southerly known occurrence. It is known around Iceland and both coasts of Greenland, as far north as Angmagssalik in the east and Upernavik in the west. In North America it is sometimes seen near southern Baffin Island and off Labrador, and is common in the Gulf of St. Lawrence and Bay of Fundy in summer, and south to North Carolina, where it can be common in winter. In the Pacific Ocean, harbor porpoises are rarely seen as far north as Point Barrow and east to the Mackenzie estuary, becoming much more common off western and southwestern Alaska. They are locally common along the entire western North American coast, from the Gulf of Alaska to Point Conception, California. They are present near islands in the Bering Sea, and along almost the entire Soviet coast, including the Sea of Okhotsk and northern Sea of Japan. Finally, they are present around the northern Japanese islands, especially Hokkaido and the northern part of Honshu.

The harbor porpoise is obviously widely distributed and still locally abundant. However, its nearshore range has made it vulnerable to heavy hunting. Native groups hunt it in many areas for food and oil, but the most destructive fisheries are those in the Sea of Azov, the Black Sea, and the Baltic Sea, where a combination of driving and netting methods have been used. The Baltic stock, once large, is now badly reduced, and the Russian fishery in the Sea of Azov and Black Sea had to be closed in the late 1960s because of depletion of cetacean stocks.

Accidental entanglement and entrapment in fishing gear is a major cause of mortality throughout the harbor porpoise's range. It dies in cod nets, salmon nets, and otter trawls in Britain; salmon nets and traps in Poland, Greenland, Newfoundland, and northwestern North America; cod traps and mackeral nets at Newfoundland; herring nets off Honshu, Japan; and white whale nets along the Kanin Peninsula, U.S.S.R. It is even caught accidentally on baited hooks set for sturgeon in the Black Sea.

Levels of organochlorine pesticides and heavy metals in harbor porpoises from northeastern North America and the North and Baltic Seas have been found to be alarmingly high.

CAN BE CONFUSED WITH: Although the taller, falcate dorsal fin of most dolphins whose ranges overlap the harbor porpoise's should make them easy to distinguish, there can be confusion under some circumstances. The rather nondescript gray coloration and inshore distribution of bottlenose dolphins may cause confusion, but their prominent beak, larger size, approachability, and dorsal-fin size and shape can distinguish them. Though it seems unlikely, a large group of harbor porpoises can be mistaken for a herd of Atlantic white-sided dolphins when viewed from the air. When harbor porpoises roll onto their sides, a flash of white on the thorax can be seen, reminding the observer of the characteristic white side of *Lagenorhynchus acutus*.

Burmeister's Porpoise

Phocoena spinipinnis
Burmeister, 1865
DERIVATION: from the Latin
spinna for "with spines," and
pinna for "fin, wing," referring
to Burmeister's observation of

spiny bumps on the leading
edge of the dorsal fin.

ZONES 4 AND 5

DISTINCTIVE FEATURES: Generally like *P. phocoena*, but with a distinctively shaped dorsal fin, fewer teeth, and some pigmentation differences; limited to temperate coastal waters of South America.

DESCRIPTION: Maximum length apparently is about 1.8 m. General body form is similar to that of *P. phocoena*, except the dorsal fin has a unique shape. It rises at a very low angle behind the middle of the back, and its rear margin is convex. It has a blunt peak. Apparently the blunt spines or tubercles sometimes present on the edge of the harbor porpoise's fin and flippers are almost always present along the front margin of this species' dorsal fin. They usually occur in three rows. The flippers are large and broad at the base, tapering to a blunt tip.

At sea the animal appears to be completely dark, although on close inspection some very weak countershading can be discerned on the ventral surface. There are 14-16 teeth in each upper jaw and 17-19 in each lower jaw, the lowest range reported for any species of *Phocoena*.

NATURAL HISTORY: The largest reported group included eight animals. Nothing is known about the reproductive biology of this porpoise, except that autumn may be a calving season. The only examined stomach contents were *Merluccius hubbsi, Pagrus sedecim,* and an unidentified squid from a specimen taken off Uruguay. Killer whales are suspected to be predators of Burmeister's porpoise.

DISTRIBUTION AND CURRENT STATUS: Burmeister's porpoise lives only in shallow temperate waters of coastal South America. On the Atlantic side its known range extends from Uruguay south to Patagonia, Argentina; on the Pacific side, from Valdivia, Chile (39°50′ S), north to Bahia de Paita, Peru (5° S). Specimens collected in recent years at Tierra del Fuego, along with a few unconfirmed sighting reports, suggest that this species may be common in portions of the Strait of Magellan and Beagle Channel. Abundance estimates have not been made, but these porpoises are believed to be much more common along the Pacific coast of South America than they are along the Atlantic coast.

The species is heavily exploited in portions of its range off Peru and Chile, both directly and incidentally. It is caught in gill nets set for sciaenid fish along the coast of Peru, and as many as 2,000 are said to have been sold at Peruvian meat markets in a single year. Chilean fishermen also catch them intentionally for bait and, perhaps, food, and accidentally in their gill nets. Smaller numbers are taken incidentally in shark gill nets in Uruguay, and at least four individuals were collected in one year in the now-defunct gill-net fishery for crabs off Chile and Argentina. The effect of these various forms of exploitation has not been evaluated.

CAN BE CONFUSED WITH: Several small odontocetes that might cause confusion share this porpoise's range. The spectacled porpoise and Commerson's dolphin should have enough conspicuous white pigmentation to make them easy to distinguish from Burmeister's porpoise. Some overlap exists with the black dolphin on the Chilean coast; this species is likely to be very difficult to distinguish from Burmeister's porpoise at sea. The franciscana is found in some of the same areas of eastern South America as Burmeister's porpoise. Its long beak should make it easy to recognize.

Cochito*

Phocoena sinus
Norris and McFarland, 1958
DERIVATION: from the Latin
sinus for "pocket, recess, or
bay," referring to the species'
distribution solely within the
Gulf of California.

ZONE 1

DISTINCTIVE FEATURES: Externally almost identical to *P. phocoena;* range limited to upper Gulf of California.

DESCRIPTION: The largest specimen measured was 1.5 m long, and it was a physically mature female. Almost nothing is known about the differences in external appearance between this species and the harbor porpoise. Because so few specimens have been examined, those differences that may exist must await better description and delineation. The body proportions, skull, and skeleton of *P. sinus* are said to be more similar to those of *P. spinipinnis* than to those of *P. phocoena.*

For specimens examined so far, 20-21 pairs of spade-shaped teeth have been counted in each upper jaw, and 18 pairs in each lower jaw.

NATURAL HISTORY: Almost nothing is known about the natural history of the cochito. A seasonal north-south movement within a very limited range has been suggested, but hardly any data are available to confirm or deny such a hypothesis. Remains of grunt, Gulf croakers, and squid were found in the stomach of one specimen.

DISTRIBUTION AND CURRENT STATUS: The known range of this porpoise is limited to the upper quarter of the Gulf of California.

The cochito has been killed accidentally in gill nets set near San Felipe and along the Sonoran coast for the totoaba, a large sea bass. This fishery began during the late 1940s, and as many as ten cochitos are known to have been caught in a single day as recently as the early 1970s. Because totoaba had declined dramatically, the fishery was closed in 1975. This may have taken some pressure off the porpoises. Nevertheless, incidental mortality in shrimp trawls and fish nets (for species other than totoaba) probably continues to occur. Reduction of productivity in the Gulf because of damming of the Colorado River and the Gulf's contamination with pesticides may also have negative effects on the cochito.

---

* Not enough data are available to produce an illustration of this species.

CAN BE CONFUSED WITH: Many common dolphins and bottle-nose dolphins live in the same area as the cochito. The comparatively low, triangular dorsal fin and blunt snout of the cochito should be adequate to distinguish it from the other two species. Also, it is more likely to be alone or in groups of two or three, and to be difficult to observe or approach. The other two species routinely approach vessels and ride bow waves, and common dolphins in the Gulf of California form large herds whose movements often create huge disturbances on the sea surface visible from great distances.

Spectacled Porpoise

Phocoena dioptrica
Lahille, 1912
DERIVATION: from the Greek
diopter for "optical instrument,
spectacles."

ZONES 3 TO 5

DISTINCTIVE FEATURES: Body similar to *P. phocoena*, but dorsal fin larger and broader at base; dorsal coloration black, except dorsal side of tail stock; ventral coloration uninterrupted white; this dichotomous arrangement sharply demarcated; found only in cold temperate and subantarctic portions of southern hemisphere.

DESCRIPTION: This species is the largest member of the genus, reaching lengths of 2.2 m. A 50-cm fetus was judged to be near term, but the size at birth is unknown. It is probably about 70 or 80 cm. The shape of this porpoise is similar to that of the harbor porpoise. The main differences are its relatively small flippers, which are rounded at the tips, and the large triangular dorsal fin, which is thought to be exaggerated in adult males and much smaller in females. The dorsal fin has a broadly rounded leading edge and a rounded peak. It is usually not falcate.

Coloration is distinctive. A line that almost bisects the sides longitudinally divides the animal strictly into an upper black half and a lower white half. The line runs just above the eye and is interrupted only at the caudal peduncle, where white or light gray sometimes extends onto the dorsal surface. The flippers range from white with a gray border to all dark. In many individuals there is a dark stripe connecting the gape and the flipper. The flukes are black dorsally, white ventrally. A black patch surrounds the eye. This patch in the type specimen was bordered by a fine white line, which gave the impression of spectacles. The lips are black.

There are approximately 18–23 pairs of spade-shaped teeth in the upper jaw, 16–19 pairs in the lower jaw.

NATURAL HISTORY: Until the mid-1970s only ten specimens had been reported. Virtually nothing is known about this species' natural history. The size and levels of development of fetuses found in July and August suggest that the season of calving might be late winter to early spring. A peculiarity of its distribution is that *P. dioptrica* is the only species in the genus known to live near offshore islands. Killer whales are likely predators of this porpoise.

DISTRIBUTION AND CURRENT STATUS: The spectacled porpoise is known from the western South Atlantic, along the South American mainland, from the southern coast of Uruguay (about 34° S) south to Tierra del Fuego. It has also been found near the Falkland Islands and South Georgia Island. Beach surveys in recent years along the Atlantic coast of Tierra del Fuego have yielded more specimens of this species (more than 100 in 1978) than of any other. It is assumed to be fairly common there, even though it is rarely seen alive in the area. A single skull found at the Auckland Islands, New Zealand, and several sightings in the New Zealand subantarctic islands and south of Tasmania are evidence of the species' presence in the Pacific Ocean. A single record exists from the Indian Ocean, from Kerguelen Island. The species may be circumpolar in subantarctic latitudes, perhaps concentrating around offshore islands as well as along continental coasts. This porpoise apparently has never been exploited directly by man, but is known to die accidentally in fishing gear.

CAN BE CONFUSED WITH: Burmeister's porpoise, a closely related species, occurs in many of the same areas as the spectacled porpoise. The black and white coloration and unique dorsal fin shape of the spectacled porpoise should make the two species fairly easy to tell apart. Also, it should be noted that of the two, only *P. dioptrica* is known to be found around offshore islands.

Dall's Porpoise

Phocoenoides dalli
(True, 1885)
DERIVATION: from the Greek
phokaina or the Latin *phocaena*
for "porpoise," and *oides* from
the Greek *eides* for "like";
William H. Dall was an
American zoologist who

provided True with "notes
upon and drawings of two
specimens . . . captured off the
coast of Alaska in 1873."

ZONES 1 AND 7

DISTINCTIVE FEATURES: Body stocky; stark black with striking white region on belly and flank; dorsal fin small, almost triangular, often partly or completely grayish white; vigorous and fast swimmer; surfacing often causes "rooster tail" of spray; found only in North Pacific, primarily north of 28° N in the eastern part and about 36° N in the western.

DESCRIPTION: Maximum body length is about 2.2 m; weight 220 kg. Sexual maturity is reached at about 1.7 m in females, about 1.9 in males. Newborn are 85 to 100 cm long.

Dall's porpoises are extremely stocky and powerfully built, with small, pointed flippers, a small, almost triangular dorsal fin, and relatively broad flukes. The head is very small (in proportion to the robust body, it appears even smaller than it is) and lacks a well-defined beak. The keel of the tail stock is sometimes very pronounced, top and bottom.

Dall's porpoises are strikingly marked. The body is black with a prominent white patch on the flanks and belly, which extends from about midbody nearly to the flukes. On the form known as True's porpoise, which is either a subspecies *P. dalli truei* or a color phase of *P. dalli,* the white begins farther forward on the abdomen, encroaches farther forward along the side, and extends farther back along the tail. The dorsal fin ranges from all black to nearly all grayish white. On most animals it is bicolored, bearing white in a triangle touching the top and the rear margin. The posterior margin of the tail flukes is fringed with a grayish white band of varying extent. Color variants, including all-black, all-white, and piebald individuals, have been reported. Dall's porpoises have a small mouth containing 19-28 teeth per row.

NATURAL HISTORY: Though aggregations, possibly feeding concentrations, of at least 200 have been reported, Dall's porpoises usually travel in groups of ten to twenty. They frequently are found in association with Pacific white-sided dolphins, from 50° N southward, and pilot whales, from about 40° N southward.

Japanese fisheries involving Dall's porpoises have enabled scientists to study their reproductive biology. Calves are born mainly in summer, from July to September. Based on a small sample of observations in the eastern North Pacific, the calving season there appears to be more protracted; some births may occur year-round. Gestation is believed to average about 11.4 months, and calves are nursed for about two years. Estimated lengths and ages at sexual maturity are 196 cm and 7.9 years for males; 186 to 187 cm and 6.8 years for females. The calving interval probably averages about three years.

Dall's porpoises are fast swimmers, whose rush to the surface usually sends a "rooster tail" of spray aloft, a feature by which they may be readily identified even when little else is visible. On a schedule of their own choosing, with disregard to area or time,

In anticipation of a split-second sojourn at the surface, this Dall's porpoise began to exhale while still submerged. (Off southern California, April 1980: Gary L. Friedrichsen.)

An unusually clear look at the dorsal side of bow-riding Dall's porpoises. (Off southern California, October 1979: Robert L. Pitman.)

Dall's porpoises charge to a moving vessel to ride bow or stern waves. Once there, they may remain in the bow wave for half an hour or more, moving very jerkily, darting about with amazing quickness and often making steep-angle turns.

Dall's porpoises frequent varied habitats, including sounds, inland passages, nearshore regions (usually near deepwater canyons), and the open sea. They are thought capable of deep diving, and consume squid, crustaceans, and such fish as saury, hake, herring, jack mackerel, and mesopelagic, bathypelagic, and deepwater benthic species. Dall's porpoises are sometimes preyed upon by killer whales and, less frequently, by sharks.

DISTRIBUTION AND CURRENT STATUS: Dall's porpoises inhabit the northern North Pacific, being present in the west north of Choshi and central-eastern Honshu, Japan; and in the east north from about 28° N. They are especially abundant in the Sea of Okhotsk and the southern Bering Sea, and at least in summer remain as far north as the Pribilof Islands.

The separation of the different stocks and details of migrations of Dall's porpoises are poorly known. In the western Pacific, differences have been noted among animals from three areas: the Pacific coast of Japan; the Sea of Japan and the Okhotsk Sea; and the offshore northwestern Pacific and western Bering Sea. True's porpoises are abundant only off the Pacific coast of northern Japan and off the Kuril Islands, where they overlap with Dall's porpoises. In the eastern Pacific and eastern Bering Sea, only a single stock is recognized. There are year-round residents over much of the range, including the waters from about 34° N to the Aleutians. Migrations into the Bering Sea, as far as the Pribilof Islands, may occur in spring through fall, when water temperatures warm to a crisp 3° C. Southern limits appear to expand with the cooling of waters below about 13° C, reaching as far as 28° N in the east in cooler years. Dall's porpoises are present year-round in oceanic zones of the western Pacific, to at least 100 km from shore. They are found nearer shore in Puget Sound, British Columbia, Alaskan inside waters, much of the Aleutians, the Kamchatka Peninsula, and Japan. Apparent inshore movement augments the population(s) off Monterey and off southern California during winter and spring. Although the population numbers are not known, Dall's porpoises appear abundant throughout their range.

During the 1960s Japanese fishermen harpooned as many as an estimated 2,500 Dall's porpoises annually for food, and killed and discarded another 10,000–20,000 that became entangled in monofilament gill nets used in the high-seas fishery for salmon. Recent estimates of the annual incidental harvest are much lower. Small numbers of Dall's porpoises similarly entangle in coastal nets of United States salmon fishermen. Though small numbers have been held captive, the species' fragility in confinement prevents extensive live captures.

CAN BE CONFUSED WITH: When they can be examined at close range, their very distinctive characteristics make it highly unlikely that Dall's porpoises will be confused with any other species. From a distance, however, they may be confused with Pacific white-sided dolphins, which sometimes also create a "rooster tail" as they surface at high speeds. Dall's porpoises have a small, triangular dorsal fin and a distinctive white patch on the chest and belly. White-sided dolphins have a tall, falcate dorsal fin and more complex coloration. White-sided dolphins are acrobatic and entertaining, but Dall's porpoises very rarely leap even part way out of the water.

Reports of groups of small killer whales are sometimes prompted by sightings of Dall's porpoises. The two species should be readily distinguishable by differences in overall size and in the size and shape of the dorsal fin.

Finless Porpoise

Neophocaena phocaenoides
(G. Cuvier, 1829)
DERIVATION: from the Greek
neo for "new," and from the
Greek *phokaina* or the Latin
phocaena for "porpoise," named
"for its close relationship to

Phocaena"; *oides* from the Greek
eides for "like."

ZONES 1 AND 3

DISTINCTIVE FEATURES: Head rounded with no beak; head can be rotated freely; dorsal fin completely absent; uniform slate gray color; small size (less than 2 m); found only in coastal waters of the Indo-Pacific.

DESCRIPTION: Maximum known length is about 1.9 m, although physical maturity is usually attained at lengths of 1.6 m or less. Males are slightly larger than females. Several calves born in the Yangtze River in early March were about 55 cm long.

The finless porpoise looks somewhat like a small beluga. Its melon is bluff and rounded but not bulbous. There is no beak. The mouthline is curved upward toward the eye. There is a slight depression behind the crescentic blowhole, which may be regarded as a neck crease. Like the beluga, this animal has great flexibility in its neck. The blowhole is on top of the head.

The finless porpoise has no dorsal fin. A conspicuous denticulated area, variable in shape and extent, begins ahead of midback and extends far back on the dorsal surface of the peduncle. The skin on this area is dark and covered with small tubercles, or horny papillae, which one early commentator speculated were "vestiges of ancestral armor." The trailing edge of the flukes is concave, and there is a notch between the flukes. The flippers are relatively long and taper distally to a blunt tip.

The name "finless black porpoise," once widely used for this species, is a misnomer, for only after death is this porpoise black. In life it is a uniform gray, often with a bluish tinge on the back and sides. The ventral surface is lighter, with whitish zones on the throat—sometimes the upper lip also—and in the anal region. The skin is said to darken slightly with age.

There are 13-22 teeth in each row, numbering at least 52 in all. Some have laterally compressed, spatulate crowns, as in the genus *Phocoena*.

NATURAL HISTORY: Concentrations of as many as fifty of these porpoises are sometimes seen near Japan. Such groups comprise subgroups numbering five to ten animals. In some areas, like the Indus delta, singles and groups of four or less are the rule during winter. In the Yangtze River groups of two to twelve are common, although concentrations of more than twenty individuals have been reported. Finless porpoises are often seen in the same area as, but not in the company of, Indo-Pacific hump-backed dolphins. They also share much of the beiji's range in the Yangtze River.

There appears to be a peak of calving in October in Japan, and between February and April in the Yangtze River. Little else is known about reproduction. Young calves are said to be carried on their mother's back, holding onto her with their flippers. These porpoises are thought to live about 25 years.

This species' habitat is primarily coastal and riverine. Along the coasts of Japan large groups appear to migrate within 5 km of shore. They are found in mangrove communities in southern Asia, and, at least in the Indus delta, are believed to move inshore in October and offshore at the end of April, apparently following concentra-

Three captive finless porpoises arch to dive. (Toba Aquarium, Japan, May 1975: Kenneth C. Balcomb.)

tions of prawns.

The finless porpoise is said to be shy and hard to approach, but this is not always true. In the Yangtze River it is apparently not frightened by passing vessels and human activities, even though it is hunted there with "fish forks." Like the harbor porpoise, the finless porpoise is not particularly demonstrative, and has even been described as sluggish compared to many dolphins. However, its swimming capabilities appear adequate for navigating in strong currents. In comparison to the sympatric beiji in the Yangtze River, the finless porpoise is active and conspicuous, sometimes leaping a meter above the surface and "standing" vertically with the head and much of the body exposed. It is well adapted to living in shallow water; e.g., in the Indus delta it is sometimes cut off at low tide in

A finless porpoise accepts a fish from its trainer. (Toba Aquarium, Japan: Warren J. Houck.)

small rivulets or pools, and must wait until the flood tide to regain open water.

The diet of finless porpoises includes small squid, shrimp, and prawns, as well as small fish. They are said to eat eggs attached to leaves and thereby to ingest some vegetable matter, perhaps unintentionally. Rice and other grains have been found in stomachs of finless porpoises.

Specimens have survived in captivity in Japan for as long as ten years.

DISTRIBUTION AND CURRENT STATUS: These porpoises are found in warm rivers and coastal waters from Pakistan, along the entire Indian subcontinent, throughout southeast Asia and Indonesia, and north to China (far up the Yangtze River, as far as Lake Dongtinghu (formerly called Tung T'ing Lake), Korea, and Japan (where they venture into Tokyo Bay, Ise Bay, and the Seto Inland Sea). There may be enough morphological differences between the Indo-Pakistan population (on the Malabar coast) and the Sino-Japanese population (in Nagasaki Harbor) that they should be considered separate races or subspecies. The northern limit of the species is probably the vicinity of Ojika Peninsula, Miyazki Prefecture, in northern Japan.

Finless porpoises frequently are in conflict with coastal people. They are hunted with guns, harpoons, and "fish forks" in China, and many become accidentally trapped in shore seines and drift- or set-nets. Small numbers have also been captured for live display in Japan.

CAN BE CONFUSED WITH: Because of its small size, finless back, and coastal haunts, this species is unlikely to be confused with any other in most portions of its range. Where it is sympatric with the Irra-waddy dolphin, which has a similar body shape and light color, the two species can readily be distinguished by the fact that the Irra-waddy dolphin has a dorsal fin, and is larger. The long beak and tri-angular dorsal "hump" of the beiji and the susu should prevent confusion with the blunt-headed finless porpoise.

9. The River Dolphins and the Franciscana

Family: Platanistidae

Indus Susu

Platanista minor
Owen, 1853
DERIVATION: from the Greek
platanistes, specifically applied
by Pliny to a "fish" in the
Ganges; from the Latin *minor*
for "less."

Ganges Susu

Platanista gangetica
(Roxburgh, 1801)
DERIVATION: *gangetica* refers to
the Ganges River, part of the
species' habitat.

ZONE 3

PLATANISTA GANGETICA

DISTINCTIVE FEATURES: Robust body with very long, narrow beak; bluff forehead; extremely broad, spatulate flippers; dorsal fin replaced by low, indistinct hump; eyes reduced to pinholes; blowhole a single longitudinal slit; distribution limited to Indus (*minor*) and Ganges-Brahmaputra-Meghna (*gangetica*) drainages of Indian subcontinent.

DESCRIPTION: The two recognized species of susu differ only in geographical distribution and skull morphology. Since we have no

Indus susus, like this female captured near Sukkur, Pakistan, in November 1968, are amazingly limber. (Steinhart Aquarium, San Francisco, 1969: California Academy of Sciences.)

evidence of differences in external appearance or natural history, we will treat them here as if they were separate populations of a single species.

Female susus grow to lengths of about 2.5 m; males, 2 to 2.1 m. Large females can weigh much more than 80 kg. At birth susus are probably 70 to 75 cm long.

The body of the "blind" river dolphin is robust, even chunky. Its most striking feature is the long, narrow beak, which can be as much as one-fifth of its body length. The tip of the beak is slightly thickened, and the mouthline curves upward at the corners. Though the head is proportionately small, the melon is bluff and prominent. The blowhole consists of a longitudinal gouge on top of the head. Since the cervical vertebrae are free, this dolphin's neck is very flexible. The susu's eyes are surely its most intriguing feature, for it is the only cetacean (or marine mammal, for that matter) with eyes that lack a crystalline lens. The optic opening is scarcely as large as a pinhole, barely large enough to allow penetration by light. Although blind for all intents and purposes, the susu is probably able to detect the direction, and perhaps intensity, of light in its environment.

The flippers are splayed widely from their rather slender origins to form broad spatulate paddles. Armbones are apparent under the taut skin, and the tips of these bones protrude, so that the outer margin of the flipper has a scalloped appearance. There is no dorsal fin; in its place is a degenerate ridge or hump. The flukes are broad and wide, with a concave rear margin and a median notch.

The susu's long beak is lined with an average of 27-33 sharply pointed teeth in each row.

NATURAL HISTORY: In its riverine habitat the susu appears in loose groups, rarely numbering more than ten animals. It is not noticeably gregarious, and often is found alone or in pairs.

Both sexes probably attain sexual maturity at a body length of 170 to 200 cm. Indus calves are said to be born mainly during early April.

The susu's movements and distribution are closely attuned to seasonal monsoons, which largely determine the areas accessible to the animals. In the dry season, the susu is usually restricted to main channels, but during the rains it disperses into swollen creeks and branches. It does not venture into the open sea and is limited upstream only by rocky barriers and shallow stretches.

Although much of the Ganges population's habitat remains intact, the Indus susu's environment has been altered drastically by agricultural and industrial man. Barrages constructed along the Indus for irrigation and hydroelectric power generation have parti-

A captive Indus susu, displaced from its normally opaque world, here shown swimming in its crystal-clear aquarium tank. (Steinhart Aquarium, San Francisco, 1969: California Academy of Sciences.)

tioned the population into several isolated stocks, and natural expansion and shrinkage of their range are no longer possible.

The susu is exceptional in the way it swims. As the only side-swimming cetacean, it cruises along the bottom, usually tilted to the left, nodding continuously. Since it almost never stops emitting echolocation-type pulses, the susu is believed to navigate and find food with the help of an extremely sophisticated biosonar system. Vision could be of little help in the muddy medium where the susu lives.

The susu's diet is thought to consist primarily of fish and crustaceans. A few specimens have been maintained for long periods in captivity, where they thrive on a daily ration of 1,000 to 1,400 g of fish.

The susu has no known natural predators. According to John Anderson, a nineteenth-century British naturalist who made a detailed study of the Ganges susu, local fishermen on the Indus were said at one time to catch dolphins in shallow water with the assistance of trained otters.

DISTRIBUTION AND CURRENT STATUS: The known present range of the Indus susu is entirely within the Pakistani provinces of Sind and Punjab, where it is legally protected. Most of the animals are found in Sind Province, along a 130-km stretch of the Indus between Sukkur and Guddu barrages. In Punjab, upstream from Guddu Barrage, the situation for the susu is desperate. A count of about 240 dolphins was made in the Sukkur-Guddu region in 1978; in 1979 the count was about 290. The Indus susu is unquestionably on the brink of extinction, but with continued effective protection by Pakistani authorities in Sind Province, the existing small population there may remain viable. Hopefully, protection will become effective in Punjab as well.

In contrast to the Indus population, the susus inhabiting the Ganges, Brahmaputra, and Meghna river systems of western India, Bangladesh, and Nepal are relatively abundant. Even in the polluted Hooghly River, which flows between the industrial sprawls of Calcutta and Howrah, the tenacious susus somehow persist. Several thousand are thought to survive in the deltaic zone, the treacherous lower reaches of the Ganges, where the network of navigable waters remains vast and complex. At the other end of the Ganges susu's range, however, problems similar to those faced by its western counterpart have begun to develop. Barrages in Nepal and India have interrupted the flow of rivers inhabited by the susu, isolating and perhaps condemning portions of the Gangetic population.

CAN BE CONFUSED WITH: Since susus apparently do not enter marine waters, they could be confused only with those small marine cetaceans that might enter the deltaic zones of the Ganges or Indus Rivers from time to time. Of these, the ones most likely to be confused with the susus are the finless porpoise and the Irrawaddy dolphin. However, the finless porpoise's complete lack of a beak and dorsal fin should make it easy to distinguish from the long-beaked susu, which has a relatively prominent ridge or hump in place of a dorsal fin. The Irrawaddy dolphin also lacks a beak, but has a bulbous melon and a small but definite dorsal fin. Humpbacked dolphins and bottlenose dolphins, which might occasionally venture coastward into portions of the susu's range, have dorsal fins sufficiently prominent to obviate confusion with susus.

Boutu

Inia geoffrensis
(de Blainville, 1817)
DERIVATION: *Inia* is a native name used for this dolphin by the Guarayo Indians of the San Miguel River in Bolivia; Geoffrey St. Hilaire (1772–1850) was a renowned French professor of natural history, and was instrumental in procuring zoological specimens from Portugal on behalf of Napoleon Bonaparte, one of which originated in the upper Amazon and became the type specimen for this species.

ZONES 2 AND 5

DISTINCTIVE FEATURES: Thickset body; broad-based dorsal hump; long, whiskered beak and prominent forehead; very small eyes; flippers long and broad near body; front teeth conical, rear teeth molariform; pale coloration of adults; distribution riverine in Amazon, Orinoco, and Beni basins.

DESCRIPTION: The taxonomic status of various stocks of *Inia* is not fully resolved, although most authorities agree that there is one species, perhaps subdivided by several regionally isolated subspecies. There is some evidence that those in the Orinoco River system of Venezuela are somewhat smaller than those in the Amazon River.

Maximum length is probably somewhat less than 3 m. Boutus are 70 to 83 cm long at birth.

The head is marked by a long, tubelike beak, with a long, straight mouthline that turns up slightly at the corners. A very distinctive feature of this beak is that it is lined with stiff, flattened hairs that are erect at their bases and droop toward their tips. They are apparently tactile organs. A prominent, bluff melon rises from the beak; its shape can apparently be changed at will (much like in the beluga), so that sometimes it is almost flat and at other times bulbous. The blowhole is a crescentic transverse slit on top of the head. Since the cervical vertebrae are unfused, a boutu can bend its neck; its head can make an angle of 90° to the body while it is looking down or to either side. Boutus often appear to be "scanning" as they swim, which suggests that much of their navigation is accomplished by echolocation.

The eyes of the boutu are reduced, as are those of other platanistids, but appear to be functional. Occasionally upon surfacing, the boutu brings the eyes clear of the water, possibly to inspect its surroundings visually.

The long, flexible flippers of *Inia* are impressive appendages, broad proximally, and tapering slowly to a pointed tip. They are used not only for steering but also partly for propulsion. There is no dorsal fin, but rather a broad-based hump about two-thirds of the way back from the tip of the snout. This hump is continuous with a prominent spinal ridge along the animal's back. The flukes are broad, concave along the rear margin, and divided by a modest notch. The rear edges of both the flukes and the pectoral flippers often have a frayed or scalloped appearance.

Boutus do not have a cleanly demarcated color pattern, but their pigmentation is distinctive nonetheless. Young individuals are mostly dark gray on the back, and light gray ventrally; older animals tend to be noticeably lighter overall, the back being pale whitish, and faintly tinged with bluish gray. The bodies of many boutus are suffused with pink, particularly on the undersides. The pigmentation of these dolphins is apparently partly a response to their environment, since those inhabiting turbid rivers are generally paler than those living in the relatively clear water of lakes.

The boutu's dentition is exceptional. It has two kinds of teeth: simple, conical ones anteriorly; heavier, molariform ones toward the rear of the mouth. The tooth count ranges from 24 to 34 per row.

NATURAL HISTORY: The boutu's ecology apparently is fine-tuned to fluctuations in the river's water level. Its primary prey species are benthic forms that scatter across the shallows at high water and concentrate at low-water seasons. Boutus are seen mostly as singles or pairs in the high-water season, but may assemble in groups of twelve to fifteen, sometimes even twenty, at low water.

In the upper Amazon, where males are said to become sexually active at lengths of about 2.2 to 2.3 m and females at about 1.7 to 1.8 m, most births occur between July and September. In the central Amazon, near Manaus, Brazil, the peak season for calving is June to August, and females here mature sexually at a length of about 1.3 m. Gestation reportedly lasts for slightly more than ten months; some females have been found to be simultaneously lactating and pregnant.

A boutu from the upper Amazon River, near Iquito, Peru. (Marineland of Florida, March 17, 1956: courtesy of David K. Caldwell.)

Observations of captive boutus indicate that these riverine animals, like many fish, orient themselves heading upstream or into a current. Compared to most marine dolphins, boutus are notoriously lethargic and slow-moving. Their swimming speed is usually deliberate, between 1.6 and 3.2 km per hour; bursts of 12 to 16 km per hour are possible. They come to the surface often, rarely staying submerged for as long as two minutes. Occasionally, an individual will leap a meter or more clear of the surface. Boutus have been known to approach boats. An apparently reliable story has been told of a wild individual that, in response to a fisherman's signal, would consistently assist him in herding and capturing fish. There is reason to believe that boutus rely largely on echolocation for finding food.

Most of the boutu's diet apparently consists of fish, which are usually seized with the front part of the beak and then maneuvered toward the gape, where they are crushed and chewed before being swallowed head first. The heavy molariform rear teeth, and the boutu's habit of masticating fish in captivity, have been taken as evidence that it feeds on armored or heavily scaled fishes, and perhaps molluscs and crustaceans, in the wild. Stomachs reportedly have contained characins, cichlids, and armored catfish. Captive adults consume 4 to 5 kg of fish per day.

Although they share habitats with the piranha and the cayman, nothing is known about natural predation on boutus. Several harpooned specimens have been brought into captivity, where they recovered their health and survived for more than a year. Some captured by methods other than harpooning lived in captivity for more than ten years. A boutu conceived and born at the Fort

Worth (Texas) Zoo died soon after birth. At least one individual has lived in captivity for more than sixteen years.

DISTRIBUTION AND CURRENT STATUS: The boutu inhabits much of the Amazon and Orinoco River drainages in central and northern South America, extending more than 2,800 km from the sea into Peru and Bolivia. It penetrates far up the Rio Negro of Brazil to near the Venezuelan border and to the Casiquiare in Venezuela; so there could be intercourse between the populations inhabiting the

A dead boutu is weighed in a South American jungle clearing. (Central Amazon, near Manaus, Brazil: Robin C. Best.)

Amazon and Orinoco drainages. Such a connection, however, has never been documented.

Perhaps even more than the main channels of major rivers, boutus inhabit tributaries and lakes, and during the rainy seasons the inundated jungles. As rains abate and water levels drop, they are sometimes stranded in ponds or lakes, where they must await the rain's return to gain access to main river channels. Boutus are entirely riverine.

Boutus have been captured for display in Venezuela, Colombia, Peru, and Brazil. Native superstition has prevented exploitation of the species in some parts of its range, although small catches for oil were formerly made in Brazil. Settlers who in recent years have come to live along the rivers inhabited by boutus are for the most part not inhibited by traditional taboos against hunting. As a consequence, increasing numbers of these dolphins are being killed.

CAN BE CONFUSED WITH: The only cetacean with which the boutu might be regularly confused is the tucuxi. The two species share much of the same habitat, and are sometimes seen simultaneously. However, they are easy to distinguish by the relatively tall, dolphinlike, falcate dorsal fin of the tucuxi, which contrasts with the low dorsal hump of the boutu. Also, the boutu's beak is much longer, and its body can be nearly twice as large as that of the diminutive tucuxi. The tucuxi is a more active animal, and routinely reveals more of the body to an observer than the slower, more sluggish boutu. Finally, the tucuxi is more gregarious, traveling in tight groups of two to 25 in which there is apparently strong social cohesion.

Beiji

Lipotes vexillifer
Miller, 1918
DERIVATION: probably from the Greek *leipo* for "left behind," referring to the restricted nature of the species' distribution; from the Latin *vexillum* for "flag, banner" and the suffi *fer* for "carry, bear," referring to a Chinese word that supposedly means "white flag" (Miller's informant told him that the Chinese likened this dolphin's dorsal fin to a flag).

ZONE 1

DISTINCTIVE FEATURES: Long, narrow beak, slightly upturned anteriorly; flippers broad, rounded at tips; dorsal fin low and in the shape of an isosceles triangle; light coloration; confined to the Yangtze River drainage of China.

DESCRIPTION: Body length of adults is between 2 and 2.5 m. Females are larger than males. Length at birth is about 70 to 80 cm.

The general appearance of this species is similar to that of other river dolphins (platanistids). The forehead descends steeply to a narrow, elongated beak, which is slightly upturned at the tip. The eyes are very small and degenerate, although they are believed to be functional. Their position is usually rostral, or high up on the sides of the head. The single, somewhat rectangular blowhole is oriented longitudinally on top of the head, slightly left of center. There is a barely noticeable dorsal neck crease.

The flippers are broad and rounded at the tips. The dorsal fin, somewhat behind the middle of the back, is in the shape of a low triangle, with sides fore and aft of roughly equal length.

The color is generally pale blue gray on the back and sides, grayish white on the undersides. The appendages are blue gray, while the sides of the head, the chin, and the lower part of the upper jaw are light. The conical teeth are uniform in size, and there are 30-35 in each row.

NATURAL HISTORY: The beiji's range has been inaccessible to western researchers; so study has been difficult. Fortunately, Chinese investigators have begun to take a great interest in the species, and during the last few years more information has become available.

Groups of two to six individuals are usual, with up to ten or twelve occasionally seen at once. Beiji are said to be slow but strong swimmers, and they rarely dive for more than one or two minutes. They are apparently difficult to approach in a motor boat, and are easily frightened. Propeller injuries are said to be a major cause of mortality.

Little is known about reproducton. Animals slightly more than 2 m in length have been found to be sexually mature. Mating reportedly occurs in April and May; judging from the size of fetuses and the timing of observations of newborns, scientists have estimated that gestation lasts about 10 months. Lactating specimens have been collected in September and December.

The beiji is not known to migrate, but its distribution is somewhat affected by seasonal changes in water levels and short-term movements of prey species.

In the past, during spring floods, it was able to enter Lake Dongtinghu (formerly transliterated as Tung T'ing Lake), but as the water level receded, it was forced back into the main channel of the Yangtze River. Much of the beiji's time is spent in deep channels; it occasionally approaches close to shore, apparently while feeding.

These animals apparently often hunt in the mud at or near the bottom. They have been seen in shallow water working up the mud in their search for food. Much of their hunting is said to be done at night or in early morning, over shallow sandbanks and in river branches and river mouths where prey is concentrated. A species of large-scaled fish and a "long, eel-like catfish" are the only known prey. No predators are known.

DISTRIBUTION AND CURRENT STATUS: This dolphin lives only in the Yangtze (Changjiang) River drainage on the Chinese mainland. Most early records come from Lake Dongtinghu, and its associated rivers, but more recent reports indicate that it is also present in Poyang Lake, the Tsien Tang River, and the middle and lower reaches of the Yangtze, at least as far as the estuary, and possibly all the way to Shanghai. It is said to be most plentiful in the lower Yangtze. The Three Gorges, above which the river valley is steep and narrow, with a swift current, is said to mark the upstream limit of the beiji's range.

At what may be the eleventh hour for the beiji, Chinese scientists have taken a keen interest in the species' biology and conservation. (At Institute of Hydrobiology, Wuhan, China: Masaharu Nishiwaki, by permission of Chen Pei-xun of the Institute.)

At the time of its discovery by Western scientists, in the early twentieth century, this dolphin was only rarely killed by local fishermen, who had a superstitious reverence for it. They did, however, use its oil for medicinal purposes and its meat for food as these became available. Traditional beliefs and attitudes have eroded enough to allow the species to be casually exploited today. Intensive fishing in the Yangtze very likely affects the potential food supply of the beiji. Also, it is documented that many are killed by being caught on fish hooks and in fish nets. Industrial development along the Yangtze and Tsien Tang has reportedly harmed the beiji, mainly by controlling water levels in rivers and lakes. For example, this dolphin now seems entirely absent from Lake Dongtinghu, where sedimentation has reduced the extent of available habitat. About six beijis were reported killed by a single explosion meant to clear reefs near the city of Yaohu.

Its population size is thought to be small, and there is an urgent need for conservation and study of this little-known dolphin. The Chinese government has accorded full legal protection to the beiji since 1975, but this will not eliminate the accidental killing or the serious consequences of intensified industrial development and resource exploitation in the Yangtze. One beiji was captured live in 1980, and it has been maintained at the Institute of Hydrobiology in Wuhan.

CAN BE CONFUSED WITH: The only dolphin sharing any portion of the beiji's range is the finless porpoise, which also occurs from Lake Dongtinghu to the mouth of the Yangtze and Shanghai. This porpoise's lack of a dorsal fin should prevent confusion.

Franciscana

Pontoporia blainvillei
(Gervais and d'Orbigny, 1844)
DERIVATION: from the Greek
pontos for "open sea, high sea,"
and *poros* for "passage, crossing";
named after Nereid, the sea
traverser, apparently from the
author's belief that this species

inhabits both fresh and marine
waters; H. M. Ducrotay de
Blainville (1777–1850) was a
celebrated French naturalist.

ZONE 5

DISTINCTIVE FEATURES: Long, extremely slender beak; broad, spatulate flippers; rounded, slightly falcate or triangular dorsal fin; total tooth count, 210–242, with 50–60 in each row; limited to temperate coastal waters of eastern South America.

DESCRIPTION: The largest measured male was about 1.6 m long; female, about 1.7 m. Females are larger than males. Length at birth is 70 to 75 cm.

This dolphin resembles in many respects the river dolphins (*Inia, Platanista,* and *Lipotes*). Its most distinctive feature is a long, slender beak with a straight mouthline that curves upward only slightly toward the rear. In adults the forehead is rounded and not particularly bluff. In young animals the forehead is bluff, and the beak proportionately much shorter. The crescentic blowhole opens anteriorly, near the center of the top of the head.

The splayed flippers have a serrated lateral margin, and the arm bones are detectable through the taut skin. The tail has a concave posterior margin and a median notch; the fluke tips are pointed. The dorsal fin, located near the center of the back, has a rounded peak, and its rear margin ranges from slightly convex to concave. Coloration is gray dorsally and paler ventrally. Young animals are brownish. There are 210–242 slender, finely pointed teeth, 50–60 or more on each side of each jaw.

NATURAL HISTORY: Individuals and groups of up to five animals are observed regularly. Larger schools have not been reported.

Sexual maturity is reached at an average body length of 131 cm in males, 140 cm in females, and at an estimated age of two to three years. Most births occur in spring and early summer (October to January). The gestation period is 10.5 months; the lactation period, eight to nine months. Mature females normally give birth every other year, and females have been examined that were simultaneously pregnant and lactating.

During the spring and summer, franciscanas reside in shallow water, inside the 20-fathom curve. They are exceptional among the platanistids in that they are not known to enter fresh water. Their diet includes a wide variety of fish, squid, octopus, and shrimp. When trapped in gill nets, franciscanas are attacked by the seven-gill shark, a small, local species, but attacks on unrestrained franciscanas by these or the larger sharks that frequent the same areas have not been documented. Killer whales, a likely predator, are common in some parts of the franciscana's range. Maximum age has been estimated as fifteen to twenty years.

DISTRIBUTION AND CURRENT STATUS: The species is confined to coastal waters of the western South Atlantic, from Valdes Peninsula in northern Chubut, Argentina (42° S), northward to the vicinity of Ubatuba, Brazil (23°30′ S). Though it is well-known in the La Plata estuary, the franciscana has never been recorded in either of

A 119-cm male franciscana caught during the late 1960s in a shark gill-net fishery. (Playa La Coronilla, Uruguay, February 1969: Robert L. Brownell, Jr.)

the two major rivers emptying into it, the Uruguay and the Parana.

Nearshore gill nets off the coasts of Argentina, Uruguay, and Brazil, set primarily for sharks, have killed many franciscanas since 1942. The estimated annual catch at Punta del Diablo, Uruguay, alone was 2,000 as of 1969. Since then this mortality is believed to have decreased markedly, because the shark fishery has moved progressively farther offshore. Although some blubber oil is sold to the tanning industry, and some liver oil is used for vitamin extraction, carcasses of captured franciscanas are usually discarded into the sea or left on the beach for foraging village pigs. The impact of this incidental gill-net entrapment on local stocks of *Pontoporia* is unknown.

CAN BE CONFUSED WITH: Several small odontocetes are present within the franciscana's range. The bottlenose dolphin is larger, with a shorter, broader beak and a higher, strongly falcate dorsal fin. The small tucuxi, which shares only the northern edge of the franciscana's range, is less easy to distinguish at sea. It is about the same size, with a similarly shaped dorsal fin and prominent beak. In living animals, the best field distinction is probably the beak: long and slender in the franciscana; well-demarcated, but much shorter and broader in the tucuxi. However, young franciscanas have short beaks, and may be almost impossible to distinguish from tucuxis. Dead animals should pose less of a problem, since the large number of teeth, and rounded, rather than tapered, flippers of the franciscana, make it easy to distinguish from the tucuxi. The spectacled porpoise shares much of the franciscana's range. Its lack of a beak and its black-and-white coloration should obviate confusion.

Selected Bibliography

We make no attempt here to list all the published and mimeographed articles, reprints, and abstracts that we consulted in preparing this book. The list of sources would be nearly as long as the text itself. But for the benefit of readers interested in learning more about cetaceans, we include here some of the books and articles we found most interesting and useful.

Andersen, H. T. *The Biology of Marine Mammals.* New York: Academic Press, 1969.

Baker, A. N. "New Zealand Whales and Dolphins," *Tuatara,* vol. 20, part 1 (1972). Biological Society, Victoria University of Wellington, New Zealand.

Berzin, A. A. *The Sperm Whale (Kashalot).* A. V. Yablokov, ed. Jerusalem: Israel Program for Scientific Translation, 1972. (First published in Russian, Moscow, 1971).

Best, P. B. "Order Cetacea," in J. Meester and H. W. Setzer, eds., *The Mammals of Africa: An Identification Manual* (Smithsonian Institution, 1971), Part 7, pp. 1-11.

Braham, H. W., W. M. Marquette, T. W. Bray, and J. S. Leatherwood, eds. "The Bowhead Whale: Whaling and Biological Research," *Marine Fisheries Review,* 42 (1980), nos. 9-10, 1-96.

Brower, K., and W. R. Curtsinger. *Wake of the Whale.* New York: Friends of the Earth, 1979.

Brownell, R. L., Jr. "Small Odontocetes of the Antarctic," *American Geographical Society Antarctic Map Folio Series,* folio 18 (1974), pages 13-19, and plates 8, 9, 11.

Cagnalaro, G., A. D. Natale, and G. Notarbartolo di Sciara. *Guida ai Cetacei Dei Mari Italiani.* Roma: Consiglio Nazionale Delle Ricerche, in press.

Caldwell, D. K., and M. C. Caldwell. *The World of the Bottle-nosed Dolphin.* Philadelphia and New York: Lippincott, 1972.

Chapman, J. A., and G. A. Feldhamer, eds. *Wild Mammals of North America: Biology, Management, and Economics.* Baltimore: The Johns Hopkins University Press, 1982.

Duguy, R., and D. Robineau. *Guide des mammifères marins d' Europe.* Neuchâtel and Paris: Delachaux and Niestlé, 1982.

ELLIS, R. *The Book of Whales.* New York: Knopf, 1980.

EVANS, W. E., ed. "The California Gray Whale," *Marine Fisheries Review*, vol. 36 (1974), no. 4, 1-64.

Food and Agriculture Organization of the United Nations (FAO). *Mammals in the Seas: Report of the FAO Advisory Committee on Marine Resources Research, Working Party on Marine Mammals.* Rome: FAO Fisheries Series, vol. 1, no. 5, 1978.

FRASER, F. C. *British Whales, Dolphins, and Porpoises: A Guide for the Identification and Reporting of Stranded Whales, Dolphins, and Porpoises on the British Coasts.* London: British Museum (Natural History), publication no. 549, 1976.

GASKIN, D. E. *Whales, Dolphins, and Seals: with Special Reference to the New Zealand Region.* London: Heinemann, 1972.

HALEY, D., ed. *Marine Mammals of Eastern North Pacific and Arctic Waters.* Seattle: Pacific Search Press, 1978.

HARRISON, R. J. *Functional Anatomy of Marine Mammals.* London: Academic Press, 3 vols., 1972, 1974, 1977.

HARRISON, R. J., AND J. E. KING. *Marine Mammals.* London: Hutchinson University Library, 1965.

HENDERSON, D. A. *Men and Whales at Scammon's Lagoon.* Los Angeles: Dawson's Book Shop, 1972.

HERMAN, L. M., ed. *Cetacean Behavior: Mechanisms and Functions.* New York: Wiley, 1980.

HERSHKOVITZ, P. *Catalog of Living Whales. Bulletin of the U.S. National Museum*, no. 246 (1966).

HOWELL, A. B. *Aquatic Mammals: Their Adaptations to Life in the Water.* Springfield: Charles C Thomas, 1930; New York: Dover reprint, 1970.

International Whaling Commission. Annual and special reports by the Commission, whose headquarters are presently in Cambridge, England, and its Scientific Committee, from 1950 on.

International Whaling Statistics. Issued annually since 1930 by the Committee for Whaling Statistics, Oslo, Norway.

KLEINENBERG, S. E., A. V. YABLOKOV, B. M. BEL'KOVICH, AND M. N. TARASEVICH. *Beluga* (Delphinapterus leucas): *Investigation Of The Species.* Jerusalem: Israel Program for Scientific Translations, 1969. (First published in Russian, 1964.)

LEATHERWOOD, S., D. K. CALDWELL, AND H. E. WINN. *Whales, Dolphins, and Porpoises of the Western North Atlantic: A Guide to their Identification.* National Oceanic and Atmospheric Administration Technical Report, NMFS circular 396, 1976.

LEATHERWOOD, S., AND R. R. REEVES. "Bottlenose Dolphins, *Tursiops truncatus* and Other Toothed Cetaceans." In Chapman and Feldhamer (1982), pp. 369-414.

LEATHERWOOD, S., R. R. REEVES, W. F. PERRIN, AND W. E. EVANS. *Whales, Dolphins, and Porpoises of the Eastern North Pacific and Adjacent Arctic Waters: A Guide to Their Identification.* National Oceanic and Atmospheric Administration Technical

Report, NMFS circular 444, 1982.

MACKINTOSH, N. A. *The Stocks of Whales*. London: Fishing News, 1965.

MACKINTOSH, N. A., AND S. G. BROWN. "Whales and Whaling," *American Geographical Society Antarctic Map Folio Series*, folio 18 (1974), pp. 2-4, and plates 1, 2, 11.

MELVILLE, H. *Moby-Dick or The Whale*. First published in 1851.

MITCHELL, E. *Porpoise, Dolphin and Small Whale Fisheries of the World: Status and Problems*. Morges, Switzerland; IUCN monograph no. 3, 1975.

MITCHELL, E. ed. *Review of Biology and Fisheries for Smaller Cetaceans. Journal of the Fisheries Research Board of Canada*, 32 (1975), no. 7, 889-983.

NORMAN, J. R., AND F. C. FRASER. *Giant Fishes, Whales and Dolphins*. London: Putnam, 1937.

NORRIS, K. S., ed. *Whales, Dolphins, and Porpoises*. Berkeley: University of California Press, 1966.

NORRIS, K. S. *The Porpoise Watcher*. New York: Norton, 1974.

Norwegian Whaling Gazette (Norsk Hvalfangst-Tidende). Issued monthly 1912-1968, by the Norwegian Whaling Association and the International Association of Whaling Companies.

PERRIN, W. F. *Variation of Spotted and Spinner Porpoise (*Genus Stenella*) in the Eastern Tropical Pacific and Hawaii. Bulletin of the Scripps Institution of Oceanography*, vol. 21 (1975).

PILLERI, G., ed. *Investigations on Cetacea*. Berne, Switzerland; 1969 on (irregular).

PRYOR, K. *Lads Before the Wind: Adventures in Porpoise Training*. New York: Harper and Row, 1975.

REEVES, R. R., AND R. L. BROWNELL, JR. "Baleen Whales: *Eubalaena glacialis* and Allies." In Chapman and Feldhamer (1982), pp. 415-444.

RICE, D. W. *A List of the Marine Mammals of the World*. National Oceanic and Atmospheric Administration Technical Report, NMFS SSRF -711, 1977.

RICE, D. W., AND A. A. WOLMAN. *The Life History and Ecology of the Gray Whale* (Eschrichtius robustus). American Society of Mammalogists, special publication no. 3, 1971.

RIDGWAY, S. H., ed. *Mammals of the Sea: Biology and Medicine*. Springfield: Charles C Thomas, 1972.

SCAMMON, C. M. *The Marine Mammals of the North-Western Coast of North America, Described and Illustrated: Together with an Account of the American Whale-Fishery*. San Francisco: J. H. Carmany. New York: G. P. Putnam's Sons, 1874.

SCHEVILL, W. E., ed. *The Whale Problem: A Status Report*. Cambridge: Harvard University Press, 1974.

SCORESBY, W., JR. *An Account of the Arctic Regions, with a History and Description of the Northern Whale Fishery*. Edinburgh: Constable. London: Robinson. Two vols. 1820.

TOMILIN, A. G. *Mammals of the U.S.S.R. and Adjacent Countries, Volume 9: Cetacea.* Jerusalem: Israel Program for Scientific Translations, 1967. (Originally published in Russian, Moscow, 1957).

WATSON, L. *Sea Guide to Whales of the World.* New York: Dutton, 1981.

Whales Research Institute. *Scientific Reports.* Issued annually since 1948 by the Whales Research Institute, Tokyo.

WINN, H. E., AND B. L. OLLA, eds. *Behavior of Marine Animals. Current Perspectives in Research, Volume 3: Cetaceans.* New York: Plenum, 1979.

WOOD, F. G. *Marine Mammals and Man: The Navy's Porpoises and Sea Lions.* Washington D.C.: Robert B. Luce, 1973.

Index

Page numbers in bold refer to illustrations.